Architecture-Based Design of Multi-Agent Systems

Danny Weyns

Architecture-Based Design of Multi-Agent Systems

Foreword by Len Bass

Danny Weyns
Katholieke Universiteit Leuven
Dept. Computer Science
Celesijnenlaan 200a
3001 Leuven
Belgium
danny.weyns@cs.kuleuven.be

ISBN 978-3-642-43998-8 ISBN 978-3-642-01064-4 (eBook)
DOI 10.1007/978-3-642-01064-4
Springer Heidelberg Dordrecht London New York

ACM Computing Classification (1998): I.2.11, D.2.11, C.3, J.7

Cover design: KuenkelLopka GmbH, Heidelberg

Printed on acid-free paper

Springer is part of Springer Science+Business Media (www.springer.com)

To Tessa, Eva, and Marina

Foreword

One of the most important things an architect can do is reflection. That is, examine systems, organizations, people and ask "What alternatives were considered and why was that particular decision made?" Thinking about the response gives an architect insight into the motivations and decision processes that others have used and this, in turn, should help the architect make better decisions in the future. A pre-requisite for doing this type of reflection is that the decisions and alternatives are made explicit. One venue that gives an architect an opportunity to do this type of reflection is during an architectural evaluation. Another venue is from a book such as this. This book lays out the design process used in building a collection of multi-agent systems.

In addition to providing a case study of a design process and the rationale for the design decisions, the topic of the book also is of great interest. Systems of the future will increasingly have the characteristics of the autonomous systems described here: they are simultaneously becoming more interconnected and more autonomous. Think of your smart phone that is mostly connected but can operate many functions while it is disconnected. These systems of the future will also increasingly operate without central control. Again, the telephone system and how cellular communication is managed provides a good example of this phenomenon.

Problems of connectivity raise issues of a node discovering that it is disconnected, other nodes discovering that a particular node is disconnected, how the node operates while it is disconnected, and reconnecting the node. The case study provides solutions to this problem in the context of autonomous vehicles within a limited geographic area. The essence of the solution provided—define the concept of a neighborhood for a node and treat neighboring nodes in a different fashion from other nodes—seems like it is more general than the particular application in the case study but that is still to be determined.

Communication and protocols seem to be basic to providing solutions to relatively autonomous nodes, and the structure of the middleware for such systems is of interest independently from the particular application area. The middleware needs to provide not only communication structure and neighborhood definition but also security and authentication services. The case study describes a middleware structure, and a portion of the reflection process of the architect is to ask not only about the rationale for the specific decisions made but also how well these decisions will generalize to other situations that the architect can envision. Designing systems is one aspect of the work of an architect but as stated in the introduction "Developing multi-agent systems software is 95% software engineering and 5% multi-agent systems theory." All of the portions of the software engineering life cycle must operate efficiently in order for systems to be effectively constructed. This means that the important requirements must be identified, a design generated, the design documented and evaluated, and the system constructed from the design. Each of these topics is treated in the book.

- The important requirements are typically the quality attribute requirements. Eliciting these requirements requires a different mindset from the normal requirements process whether through a formal process, through user stories, or through some other technique. Quality attribute requirements tend to be the requirements that are taken for granted by the user until they have not been met.
- Documentation is important for helping the designer think through difficult design issues and for communicating the design to others. From the perspective of a reflecting architect, the documentation provided in this book provides some understanding of the division of functionality and design rationale.
- Evaluation of a design is an important step for verifying nothing has been missed by the architect. An evaluation is an application of the multiple eyes principle—get an outside, knowledgeable perspective to look at the design. As was pointed out, this process takes time. Partially this is because it takes time to educate the outside eyes and partially because evaluation requirements look at the design with a variety of different concerns. In the ATAM process, these concerns are expressed as scenarios but, in general, looking at a complicated design in sufficient detail to determine potential problems will take time. This time could be done as one activity when the outside eyes have to travel, as in the ATAM, or the time could be spread over a collection of shorter activities when the outside eyes are generally available.

In summary, this book is interesting both for its expressed topic—the design of multi-agent systems—and as a case study where a reader can read, and reflect on, the rationale for the approach taken in building such a system.

Pittsburgh, Pennsylvania, USA

Len Bass
October 23, 2009

Acknowledgements

This book is based on 8 years of applied research conducted by a team of researchers at DistriNet Labs in collaboration with multiple industrial partners. First and foremost, I must acknowledge the contributions of Tom Holvoet and Kurt Schelfthout. Their insights and stimulating discussions have significantly contributed to the development of architecture-based design of multi-agent systems. Kurt invented the ObjectPlaces middleware described in Chapter 5. The following people must also be acknowledged for their invaluable contributions to the approach: Alexander Hellboogh, Nelis Boucké, and Elke Steegmans. The advanced coordination mechanisms described in Chapter 6 are the result of an enjoyable collaborative effort with Nelis over multiple years. I thank my former colleagues Koen Mertens and Tom De Wolf for their contributions. I also thank the Master students that have supported the development of various aspects of the approach described in this book, in particular Robin Custers, Olivier Glorieux, Bart Demarsin, Wannes Schols, Els Helsen, and Koen Deschacht. I am grateful to Jan Wielemans, Tom Lefever, Walter De Feyter, Rudy Vanhoutte, Wim Van Betsbrugge, Jan Peirsman, Raf Sempels, and Jan Vercammen for the collaboration in the EMC2 project. I wish to acknowledge the international colleagues I worked with during the past years for the stimulating collaborations. I am particularly grateful to Van Parunak and Fabien Michel for the enjoyable joint efforts. I would like to thank the researchers and architects of the Software Engineering Institute for the inspiring discussions on the design and evaluation of decentralized architectures. Several anonymous reviewers commented on earlier versions of this manuscript and suggested many improvements. Ralf Gerstner provided useful advice toward getting this book published. Thank you.

Danny Weyns

Contents

Acronyms

ADD	Attribute-Driven Design
ADL	Architecture Description Language
AGV	Automatic Guided Vehicle
APL	Agent Programming Language
ATAM®	Architecture Tradeoff Analysis Method®
AUML	Agent Unified Modeling Language
BDI	Belief, Desire, Intention
C&C	Component and Connector
cfp	call for proposals
CNET	Contract NET
CORBA	Common Object Request Broker Architecture
CPU	Central Processing Unit
DynCNET	Dynamic Contract NET
E'nsor®	Egemin navigation system on robot
E'pia®	Egemin platform for integrated automation
ERP	Enterprise Resource Planning
FiTA	Field-based Transport Assignment
IEC	International Electrotechnical Commission
IEEE	Institute of Electrical and Electronics Engineers
IP	Internet Protocol
ISO	International Standard Organization
JEE	Java Enterprise Edition
LAN	Local Area Network
LIME	Linda In a Mobile Environment
Mbps	Megabits per second
QAW	Quality Attribute Workshop
RMI	Remote Method Invocation
SOAP	Simple Object Access Protocol
TB	Transport Base
UML	Unified Modeling Language
XML	Extensible Markup Language

Chapter 1
Introduction

A well-known claim for multi-agent systems is that they are especially suited to develop software systems that are decentralized, can deal flexibly with dynamic conditions, and are open to system components that come and go. While we endorse this claim, developing real-world multi-agent systems taught us that achieving these goals is a complex engineering problem. Our experience with real-world multi-agent systems development can be captured succinctly in the following statement:

> Developing multi-agent systems software is 95% software engineering and 5% multi-agent systems theory.

In this book, we present *architecture-based design of multi-agent systems*, an architecture-centric approach for developing real-world multi-agent systems. The approach integrates multi-agent system concepts with state-of-the-art principles and methods from mainstream software architecture and middleware. The practical applicability of the approach is demonstrated for an industry-strength application in the domain of automated transportation systems.

The objective of the book is twofold. On the one hand, we provide a guide to software engineers for the architectural design of real-world multi-agent systems. On the other hand, we give a detailed description of how we have used this guide for developing a complex multi-agent system in an industrial setting.

We start by introducing two fields of software engineering that are central in architecture-based design of multi-agent systems: software architecture and middleware. Next, we explain how and why our perspective on engineering multi-agent systems differs from existing agent-oriented methodologies. Then, we introduce the automated transportation system where we use a case study to demonstrate the applicability of architecture-based design of multi-agent systems. The introduction concludes with an overview of the book.

1.1 Software Architecture and Middleware

Two fields of software engineering are central in this book: software architecture and middleware. In this section, we introduce both fields and illustrate their importance with respect to the design of real-world multi-agent systems.

D. Weyns, *Architecture-Based Design of Multi-Agent Systems*,
DOI 10.1007/978-3-642-01064-4_1, © Springer-Verlag Berlin Heidelberg 2010

1.1.1 Software Architecture

Since the mid-1990s, software architecture has been the subject of increasing interest in software engineering research and practice. Software architecture is a corner stone for ensuring that systems achieve their quality and functional goals and ultimately serve an organization's business and mission goals. Software architecture provides the required level of abstraction and generality to deal with the increasing challenges of adaptation required in distributed software applications [91]. Bass, Clements, and Kazman define software architecture as "the structure or structures of the system, which comprise software elements, the externally visible properties of those elements, and the relationships among them" [21]. Software elements provide the functionality of the system, while the required quality attributes (performance, adaptability, usability, modifiability, etc.) are primarily achieved through the structures of the software architecture. Software architecture constitutes a model for how a system is structured and works. This model is transferable to other systems with similar requirements and can promote large-scale reuse of design. Besides technical value, software architecture is also considered as a key vehicle for communication among stakeholders. Software architecture provides a basis for creating mutual understanding and consensus about the software system [46].

During architectural design, architects apply proven architectural approaches to transfer the system requirements into appropriate software structures. Architectural patterns offer well-established solutions to architectural problems. An architectural pattern expresses a fundamental structural organization schema for a software system which exhibits known quality attributes. For example, layers is a well-known pattern that structures a software system into an appropriate number of layers and places them on top of each other. The services of each layer implement a strategy for combining the services of the layer below in a meaningful way. Layers enhance maintainability, extensibility, and reusability of the system. However, applying the layer pattern can be expensive on system resources affecting performance.

A multi-agent system is in essence a system that is structured as a set of autonomous agents that are able to flexibly adapt their behavior to changing operating conditions. Durfee and Lesser define a multi-agent system as "a loosely coupled network of problem solvers (agents) that interact to solve problems that are beyond the individual capabilities or knowledge of each problem solver " [52]. Characteristics of multi-agent systems are as follows: (1) each agent has incomplete information or capabilities for solving the problem and, thus, has a limited viewpoint; (2) there is no system global control; (3) data is distributed; and (4) computation is asynchronous. Multi-agent systems are characterized by specific intra-agent and inter-agent structures. At the level of individual agents, many different architectures have been developed, ranging from simple reactive agents to complex reasoning agents. At the system level, the multi-agent system can be structured as an organization of selfish agents that play different roles in the organization pursuing their own interests. Other multi-agent systems consist of cooperative agents that aim to achieve a common goal. Agents can interact in different ways: via a high-level communication language and specific interaction protocols or via manipulating marks in a shared coordination medium. Since specific multi-agent system structures imbue the software systems with certain qualities, while making certain tradeoffs, a primary focus of multi-agent system engineering is on the software architecture of the system. Multi-agent systems are known for quality attributes such as adaptability, openness, robustness, and scalability, making multi-agent systems particularly interesting to deal with the demanding challenges of complex distributed software applications.

In Chap. 3, we explain how design expertise in multi-agent systems can be captured as architectural patterns. We present a number of architectural patterns for a family of multi-agent systems. These patterns embody a set of architectural best practices derived from the experiences with developing various multi-agent system applications. In Chap. 4, we explain how architectural patterns play a key role in transferring stakeholder requirements into appropriate software structures. The primary structures of a multi-agent system are critical for the achievement of the system's quality attributes. Chapter 7 elaborates on the evaluation of multi-agent system architectures. Architectural evaluation allows determining the tradeoffs and risks of architectural decisions with respect to satisfying important quality attribute requirements.

1.1.2 Middleware

Middleware is the software layer that lies between the operating system and the application components. Middleware provides high-level abstractions to support the coordination of distributed software components. With networks becoming increasingly pervasive, middleware appears as a major building block for the development of complex distributed software systems [77]. Since multi-agent systems are particularly useful for problem domains characterized by highly dynamic operating conditions and inherent distribution of resources, it is clear that any multi-agent system application should deal with the distribution concern.

Domain-specific middleware for multi-agent systems typically consider agent communication as the prior means for agent coordination. A communication infrastructure usually provides a management system that enables agents to register and locate one another and a message transport system. Coordination infrastructures offer an alternative for direct message exchange allowing agents to interact indirectly via a shared medium. Two different examples of coordination infrastructures are an electronic institution that acts as a governor for interaction and digital pheromones that agents use to mark dynamic paths to areas of interest similar as social ants. Since interactions necessary for coordination often take place in a concurrent and distributed environment that is unreliable, middleware is a crucial aspect in software development of multi-agent systems. Concerns such as security, persistency, and transactional behavior are typically supported by domain-independent middleware services. Since these concerns often crosscut the system, vertical integration with domain-specific middleware is an important aspect of the design of any real-world multi-agent system.

In Chap. 5, we elaborate on the role of middleware for supporting the development of distributed multi-agent systems. We take a closer look at the multiple layers of middleware in distributed software systems in general, and we zoom in on middleware for multi-agent systems. We explain in detail the middleware used in the case study, called ObjectPlaces. ObjectPlaces proposes two new programming abstractions, *view* and *coordination role*, to support the development of mobile multi-agent system applications.

The first abstraction, a view, is an automatically up-to-date collection of data objects that are copies or representations of data objects available on a set of nodes in the network. The middleware automates gathering the data objects from a set of nodes and maintains the view in the face of dynamically changing availability of the data objects. The second abstraction, a coordination role, encapsulates the behavior of a component of the application engaging in a protocol. The middleware automates the setup and maintenance of an interaction session between a number of participating components in the mobile network, in the face of a frequently changing number of participants.

The ObjectPlaces middleware encapsulates the tedious management tasks associated with distribution in mobile multi-agent systems. The middleware has significantly reduced the complexity of tackling distributed coordination problems in the case study. In Chaps. 5 and 6 we show how the middleware has simplified the development of the application components in the automated transportation system for collision avoidance and task assignment, respectively.

1.2 Agent-Oriented Methodologies

Since the early 1990s the idea that multi-agent systems are a radically new way of engineering software has dominated research and practice in agent-oriented software engineering. Wooldridge et al. [177] state that

> There is a fundamental mismatch between the concepts used by object-oriented developers and other mainstream software engineering paradigms, and the agent-oriented view. [...] Existing software development techniques are unsuitable to realize the potential of agents as a software engineering paradigm.

Zambonelli and Omicini [182] state that

> Agent-based computing can be considered as a new general-purpose paradigm for software development, which tends to radically influence the way a software system is conceived and developed.

This vision has led to the development of numerous multi-agent system methodologies. Some of the methodologies focus on particular phases of the software development process, e.g., Gaia [177, 181]. Others cover the full software development life cycle, e.g., Tropos [64]. Some of the proposed methodologies adopt mechanisms and practices from mainstream software engineering. Prometheus [119] is inspired by object-oriented mechanisms. MaSE [174] uses practices of the Unified Process. Adelfe [25] uses constructs of the Unified Modeling Language. However, nearly all methodologies take an independent position, barely embedded in mainstream software engineering practice. Studying literature reveals that very limited results have been obtained in the application of these methodologies to real-world problems. A notable exception is the DACS methodology (Designing Agent-based Control Systems), introduced by Bussmann et al. [38], that was applied in the design of a multi-agent system for manufacturing control at DaimlerChrysler.

Our perspective on engineering multi-agent systems starts from the viewpoint that multi-agent system engineering fits well within mainstream software engineering. This vision is based on the observation that multi-agent systems are essentially a way to structure a software system, making software architecture of paramount importance in engineering multi-agent systems.

By putting software architecture and middleware at the heart of the engineering process, architecture-based design of multi-agent systems places multi-agent systems in a larger context of software engineering. This perspective provides at the same time insights and opportunities for both multi-agent system and mainstream software engineers and researchers. Considering multi-agent systems from a software architecture perspective does not delude existing results of agent-oriented software engineering. On the contrary, agent-oriented software engineering has developed a wide body of valuable concepts and techniques for engineering agent behavior, adaptation, advanced interactions, organizations, learning, etc. This domain-specific know-how is required to support architects and developers of multi-agent systems. The architecture-based approach for developing multi-agent systems presented in this book integrates such domain-specific concepts and techniques with mainstream software engineering methods and practices.

1.3 Case Study

Throughout this book, we use an automated transportation system as a case study. The description of the architectural design and development of this application demonstrate the practical applicability of architecture-based design of multi-agent systems. The case study was developed between 2004 and 2007 by a team of engineers and developers of Egemin, a leading company that provides full life cycle support for automated transportation systems [53], and researchers of DistriNet Labs. This section introduces the application and motivates the use of a multi-agent system architecture.

An automated transportation system consists of a number of automatic guided vehicles (AGVs) that need to work together to transport loads in an industrial environment. Transports are generated by client systems, typically an enterprise resource planning (ERP) system. The main functionalities that an AGV transportation system has to fulfill are assigning incoming transport tasks to an appropriate AGV, routing the AGVs through the warehouse efficiently while avoiding collisions and deadlocks, and recharging the AGVs' batteries.

An AGV transportation system has to deal with dynamic and changing operating conditions. The stream of transports that enter the system is typically irregular and unpredictable, AGVs can leave and re-enter the system for maintenance, production machines may have variable waiting times, etc. All kinds of disturbances can occur, supply of goods can be delayed, certain areas in the warehouse may temporarily be closed for maintenance services, loads can block paths, AGVs can fail, etc. Despite these challenging operating conditions, the system is expected to operate efficiently and robustly.

Traditionally, the AGV systems deployed by Egemin are directly controlled by a central server. The server plans the schedule for the system as a whole, dispatches commands to the AGVs, and continually polls their status. This results in reliable and predicable solutions. The central point of control also enables easier diagnosis of errors. However, a shift in user requirements challenges the centralized architecture. Customers increasingly request for *self-managing systems*, i.e., systems that are able to adapt their behavior with changing circumstances autonomously. Self-management with respect to system dynamics translates to two specific quality requirements: flexibility and openness. Flexibility refers to the system's ability to deal with dynamic operating conditions. Openness refers to the system's ability to deal with AGVs leaving and entering the system.

To deal with these new quality requirements, a radically new architecture was conceived based on situated multi-agent systems. Applying a situated multi-agent system opens perspectives to improve flexibility and openness of the system: the AGVs can adapt themselves to the current situation in their vicinity, order assignment is dynamic, the system can deal autonomously with AGVs leaving and re-entering the system, etc. However, introducing a decentralized architecture may have a number of implications, in particular decentralized decision making may have an impact on the overall efficiency of the system such as throughput and bandwidth usage. These are critical issues that have to be considered during the design and development of the multi-agent system.

The software system was implemented on a prototype setup with real AGVs and tested in larger, industrially used simulations. The design and implementation of the AGV control system needed 8+ man-years of effort. The delivered code base for the control software consists of about 100K lines of C# code. This system interfaces with a lower level AGV steering system that for its real-time properties is written in C.

1.4 Overview of the Book

In Chap. 2, we give a general overview of architecture-based design of multi-agent systems. We situate architectural design in a software development life cycle, and we zoom in on the different steps in the approach. These steps include requirements elicitation, architectural design, architecture documentation, and architecture evaluation. For each step, we give some background and we introduce the different techniques and methods that are used. The chapter concludes with a brief explanation of how software architecture serves as a basis for downstream design and implementation of the system.

Chapter 3 shows how architectural patterns provide the means to capture expertise with multi-agent system design. We introduce a set of architectural patterns for situated multi-agent systems, the family of multi-agent systems we have applied in the case study. The set of architectural patterns provides an asset base that architects can use in the design of a family of multi-agent systems.

In Chap. 4, we elaborate on architectural design of multi-agent systems and the documentation of software architecture. In architecture-based design of multi-agent systems, we use attribute-driven design (ADD) [173] as a design method. ADD is concerned with the high-level decomposition of a software system which is critical for satisfying the system's quality requirements. ADD yields a set of architectural views. To document the views we use the Views and Beyond [45] method. We apply the methods to the design and documentation of the case study. The case study makes clear how the various patterns for situated multi-agent systems were applied during architectural design.

In Chap. 5, we zoom in on middleware for distributed multi-agent systems. Middleware supports application developers with the design and implementation of coordination solutions in a distributed setting. We explain in detail a concrete middleware that was developed for the case study and we illustrate how this middleware supported a complex coordination problem in a mobile setting.

One particularly complex coordination problem in distributed multi-agent systems is task assignment. Chapter 6 is dedicated to this problem. We zoom in on two approaches for adaptive task assignment that are characteristic for two classical families of coordination mechanisms for task assignment: a protocol-based approach and a field-based approach. We explain the design and validation of both approaches in the case study, and we make a tradeoff analysis.

Chapter 7 elaborates on the evaluation of a multi-agent system architecture. In architecture-based design of multi-agent systems, we use the Architecture Tradeoff Analysis Method (ATAM) [46][1] for the evaluation of software architecture. ATAM is a structured method to examine whether a software architecture is suitable for the system for which it was designed. ATAM uncovers architectural tradeoffs and risks in the design. We explain in detail the ATAM evaluation for the case study and reflect on the experiences with using ATAM for the evaluation of a multi-agent system architecture.

In Chap. 8, we discuss related approaches that explicitly connect software architecture with multi-agent systems. We also examine related work on middleware for mobile systems. Additionally, we give a brief overview of related work on the control of AGV transportation systems.

In Chap. 9, we reflect on architecture-based design of multi-agent systems and its application to the case study, and we report lessons learned from applying the approach in practice. We conclude with an outline of challenges for future research on engineering multi-agent system derived from our experiences.

[1] Architecture Tradeoff Analysis Method® and ATAM® are registered in the U.S. Patent and Trademark Office by Carnegie Mellon University.

Chapter 2
Overview of Architecture-Based Design of Multi-Agent Systems

Architecture-based design of multi-agent systems puts software architecture at the center of the software development activities. In this chapter, we give an overview of the approach. We start by situating architecture-based design in a software development life cycle and we give an overview of the methods used in the different steps of architecture-based design of multi-agent systems. Next, we zoom in on the different steps in the approach, including requirements elicitation, architectural design, architecture documentation, and evaluation. For each step, we give the necessary background and we introduce the different techniques and methods that are used. We conclude with a brief explanation of how software architecture serves as a blueprint for system development and a summary.

2.1 General Overview of the Approach

To understand the approach of architecture-based design of multi-agent systems, we first situate architectural design in a software development life cycle. Then, we give an overview of the techniques and methods that are used in the different steps of architecture-based design of multi-agent systems.

2.1.1 Architectural Design in the Development Life Cycle

We use the evolutionary delivering life cycle [108, 21], see Fig. 2.1. This life cycle model puts architectural design in the middle of the development activities. The main idea of the model is to support incremental software development and to incorporate early feedback from the stakeholders. The life cycle consists of two main phases: developing the core system and delivering the final software product. Our focus is on architectural design and its connecting activities.

In the first phase the core system is developed. This phase includes four activities: defining a domain model, performing a system requirements analysis, designing the software architecture, and developing the core system. Defining the domain model is documenting a vocabulary of the key concepts and their relationships

D. Weyns, *Architecture-Based Design of Multi-Agent Systems*,
DOI 10.1007/978-3-642-01064-4_2, © Springer-Verlag Berlin Heidelberg 2010

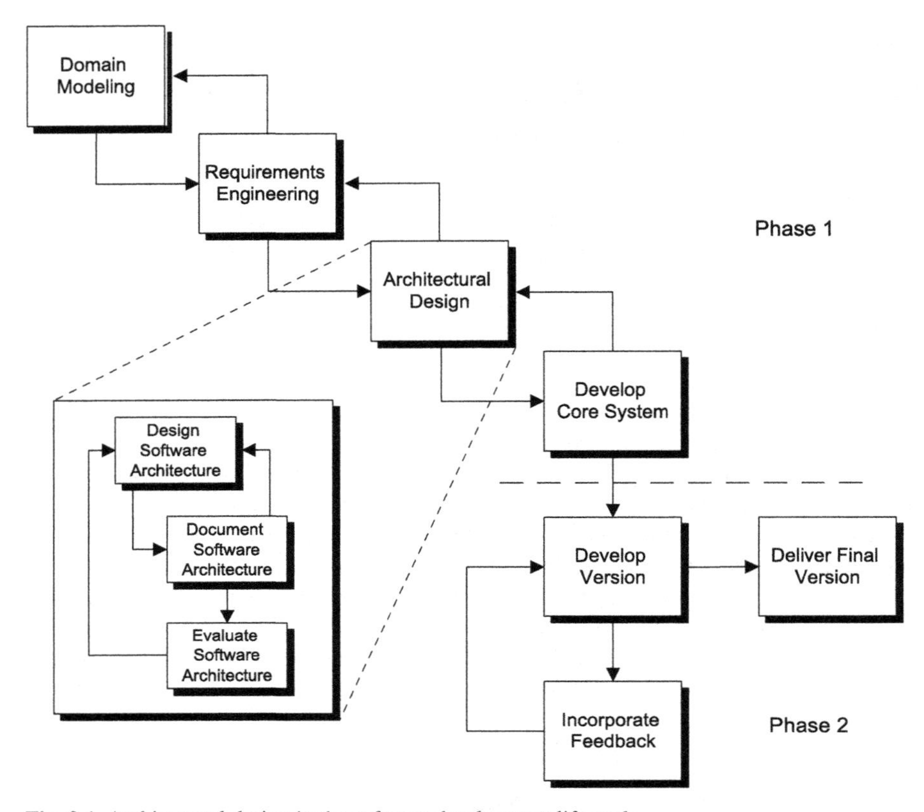

Fig. 2.1 Architectural design in the software development life cycle

of the domain of the system being developed. Requirements analysis includes the formulation of functional requirements of the system as well as eliciting and prioritizing of the quality attributes requirements. Designing the software architecture includes the design and documentation of the software architecture and an evaluation of the architecture. The development of the core system includes downstream design, implementation, and testing. The software engineering process is an iterative process, the core system is developed incrementally, passing multiple times through the different stages of the development process. Figure 2.2 shows how architectural design iterates with requirements analysis on the one hand and with the development of the core system on the other hand. The output of the first phase is a domain model, a list of system requirements, a software architecture, and an implementation of the core of the software system.

In the second phase, subsequent versions of the system are developed until the final software product can be delivered. In principle there is no feedback loop from the second to the first phase, although in practice specific architectural refinements may be necessary.

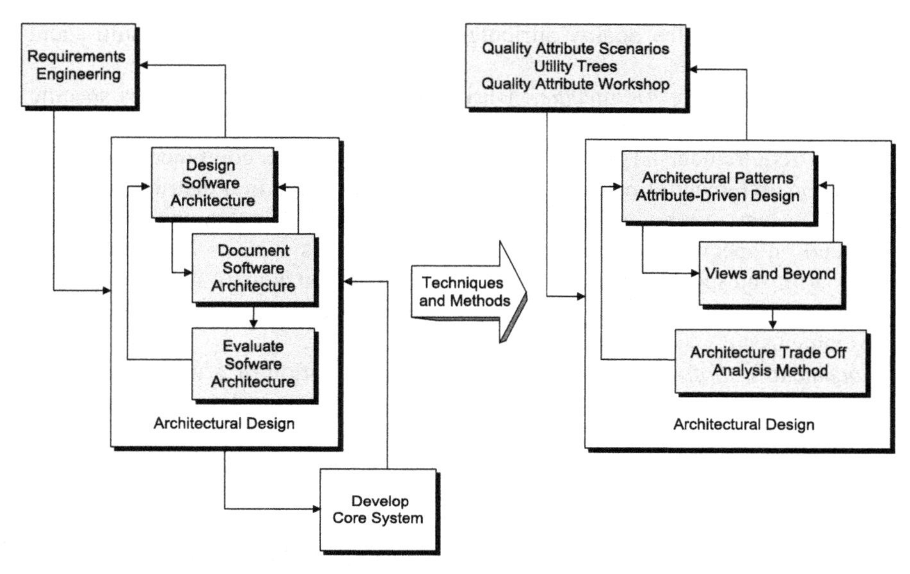

Fig. 2.2 On the *left hand side*: The steps of architecture-based design of multi-agent systems in the life cycle. On the *right hand side*: The mapping of techniques and methods on each of the steps. *Shaded boxes* represent the main activities of interest in the approach

The focus of architecture-based design of multi-agent systems is on the first phase in the life cycle.

2.1.2 Steps of Architecture-Based Design of Multi-Agent Systems

Now, we give a bird's-eye view on the activities of architecture-based design of multi-agent systems and we map the techniques and methods we use on the different steps of the approach.

Several of the methods we use in architecture-based design of multi-agent systems are developed at the Software Engineering Institute [3] of Carnegie Mellon University, including the quality attribute workshop, attribute-driven design, and the Architecture Tradeoff Analysis Method. These methods have proven their value in mainstream software engineering practice. In architecture-based design of multi-agent systems, the methods are scoped toward the domain of multi-agent systems. This scoping is reflected in each step in the approach:

- *Requirements Engineering*. The elicitation and specification of quality attribute requirements focuses on attributes which are particularly relevant for the domain of multi-agent systems. Examples are adaptability, openness, and scalability. The decision to apply a multi-agent system architecture should be based on a good understanding of the relative importance between the regular quality attribute

requirements and the quality attributes that can be achieved by a multi-agent system architecture.
- *Designing Software Architecture.* During architectural design, patterns specific to the domain of multi-agent systems are employed to achieve the stakeholders' quality requirements. Typically, these patterns have to be combined with other common architectural patterns. Support for coordination in a distributed multi-agent system requires a suitable middleware. Middleware requirements may depend on specific properties of multi-agent systems such as decentralization of control and specific characteristics of the application domain such as mobility. Middleware can have a severe impact on quality attributes such as efficiency and resource usage.
- *Documenting Software Architecture.* The documentation of a multi-agent system architecture includes views and models that are typical for the domain of multi-agent systems. Examples are models for describing protocols in high-level agent communication languages, models to describe the roles and dynamics of agent organizations, etc. The description of such models may require dedicated modeling languages.
- *Evaluation of Software Architecture.* The evaluation of multi-agent system architectures includes the evaluation of architecture approaches specific to the domain of multi-agent systems, in particular decentralization of control. Architecture evaluation allows to pinpoint not only the advantages of a multi-agent system architecture, but also the tradeoffs and risks implied by the decentralized architecture.

In the remainder of this chapter, we zoom in on the different steps of the approach. In Sect. 2.2, we discuss requirements eliciting, the preparatory step to start architectural design. Next, we discuss architectural design in Sect. 2.3. In this step, the various system requirements are achieved by selecting suitable architectural patterns and assigning responsibilities to the various architectural elements. We briefly discuss middleware support for multi-agent systems in Sect. 2.4. Then, we explain how a software architecture is documented in Sect. 2.5, and we zoom in on the evaluation of software architecture in Sect. 2.6. Finally, in Sect. 2.7 we briefly explain how software architecture serves as a blueprint for downstream design and implementation of the system. For each step, we give the necessary background and we introduce the different techniques and methods that are used.

2.2 Functional and Quality Attribute Requirements

Architectural design can start when the most important system requirements are known. This set of requirements is usually called the architectural drivers and includes functional and quality requirements.

Functionality is the ability of the system to perform the tasks for which it is intended. To perform a task, software elements have to be assigned correct responsibilities for coordinating with other elements to offer the required functionality.

Functional requirements of a system are typically expressed as use cases, see, e.g., [96]. A use case lists the steps necessary to accomplish a functional goal for an actor that uses the system. We also use scenarios that describe interactions among parts in the system—rather than interactions that are initiated by an external actor. Consider as an example a use case that describes the requirement of collision avoidance of AGVs on crossroads:

> The goal of the use case is to prevent AGVs from colliding at crossroads. When two or more AGVs approach a crossroad simultaneously, the control system should allow only one vehicle at a time to pass the crossroad.

Functionality does not depend on the structure of the system. In principle, if functionality were the only requirement, the system could exist as a single monolithic module with no internal structure at all [21].

Quality is the degree to which a system meets the nonfunctional requirements in the context of the required functionality. Quality attributes are nonfunctional properties of a software system such as performance, usability, and modifiability. Achieving quality attributes must be considered throughout the development process of a software system. However, the software architecture is critical to the realization of most quality attributes; it provides the basis for achieving quality. For the expression of quality requirements we use *quality attribute scenarios* [20]. A quality attribute scenario consists of three parts:

1. Stimulus: an internally or externally generated condition that affects (a part of) the system and that needs to be considered when it arrives at the system; e.g., a user invokes a function, a component fails, a maintainer makes a change.
2. Environment: the conditions under which the stimulus occurs; e.g., at runtime with system in normal operation, at design time.
3. Response: the activity that is undertaken—through the architecture—when the stimulus arrives. The response should be measurable so that the requirement can be tested; e.g., the system switches to save mode, the error is displayed within 5 s, the change requires a person-month of work.

Here is an example of a quality attribute scenario:

> An AGV gets broken and blocks a path under normal system operation. Other AGVs have to record this, choose an alternative route—if available—and continue their work.

The stimulus in this example is "An AGV gets broken and blocks a path," the environment is "normal system operation," and the response is "other AGVs have to record this, choose an alternative route—if available—and continue their work."

Quality attribute scenarios provide a means to transform vaguely formulated qualities such as "the system shall be modifiable" or "the system shall exhibit acceptable flexibility" into concrete expressions. Scenarios are useful in understanding a system's qualities; formulating scenarios help stakeholders to express their preferences about the system in a clear way. Scenarios help the architect to make directed decisions and are a primary vehicle for analysis and evaluation of the software architecture.

Ideally, the quality attribute scenarios of the system are collected and prioritized before the start of architectural design. In architecture-based design of multi-agent systems, we use a *quality attribute workshop* (QAW) [19] to elicit and prioritize quality attributes. A QAW is a facilitated method that engages stakeholders to discover the driving quality attributes of a software-intensive system. During a QAW, quality attribute scenarios are generated, prioritized, and refined. *Utility trees* [46] are one way to prioritize quality attribute scenarios. A utility tree provides a mechanism for the architect and the other stakeholders involved in a system to define and prioritize the relevant quality requirements precisely. We elaborate on utility trees in Sect. 2.6 when we discuss the evaluation of software architecture. The results of a QAW include a list of architectural drivers, a prioritized list of raw scenarios, and the refined scenarios. The architect can use this information to design the architecture. In addition, after the architecture is created, the scenarios can be used as part of a software architecture evaluation. In practice, often a number of iterations will be necessary to gather and order system's requirements.

A rigorous specification of quality attribute scenarios is key for delineating a convincing motivation for applying a multi-agent system architecture. Pinpointing the quality attributes that are typically associated with multi-agent systems and identifying conflicts with other quality attributes will help to clarify the added value and tradeoffs of adopting a multi-agent system.

We elaborate on requirements elicitation when we discuss architectural design in Chap. 4 and architecture evaluation in Chap. 7.

2.3 Architectural Design

Designing a software architecture is about moving from system requirements to architectural decisions. Besides thorough knowledge and experiences from architects, this crucial engineering step requires a well-founded design method. In architecture-based design of multi-agent systems, we use the attribute-driven design (ADD) [21, 173] method. ADD deals with high-level design of the architecture and as such can be viewed as an extension of most other development processes, such as the Rational Unified Process [93]. ADD is an iterative decomposition method that is based on understanding how to achieve quality goals through proven architectural approaches, in particular architectural patterns.

2.3.1 Architectural Patterns

Central in ADD is the achievement of a system's quality attributes based on design decisions. Such decisions are called *tactics*. A tactic is a widely used architectural approach that has proven to be useful to achieve a particular quality [21, 143]. For example, "rollback" is a tactic to recover from a failure aiming to increase availability, or "concurrency" is a tactic to manage resource access aiming to improve performance. Actually, to realize one or more tactics an architect typically chooses an appropriate *architectural pattern* [148].[1] Bass and colleagues define an architectural pattern as "a description of architectural elements and relation types together with a set of constraints on how they may be used" [21]. An architectural pattern is a recurring architectural approach that exhibits particular quality attributes. A pattern documents not only how a solution solves a problem but also why it is solved, i.e., the rationale behind this particular solution [16]. Examples of common architectural patterns are layers, pipe-and-filter, and blackboard.

Architectural patterns also provide the means to document and mature knowledge and practices with multi-agent systems. In the course of designing and building multi-agent systems, architectural patterns can be derived that provide generic solution schemes for recurring design problems. As an illustration, we briefly explain the subsumption architecture developed by Brooks [34].

The subsumption architecture describes an architectural pattern for the decision making of a single robot. The architecture is organized as a series of parallel working layers, each layer is responsible for a specific behavior of the agent. The priority of layers (behaviors) increases from bottom to top. Higher layers are able to inhibit lower layers, giving priority to more important behavior. Figure 2.3 shows an

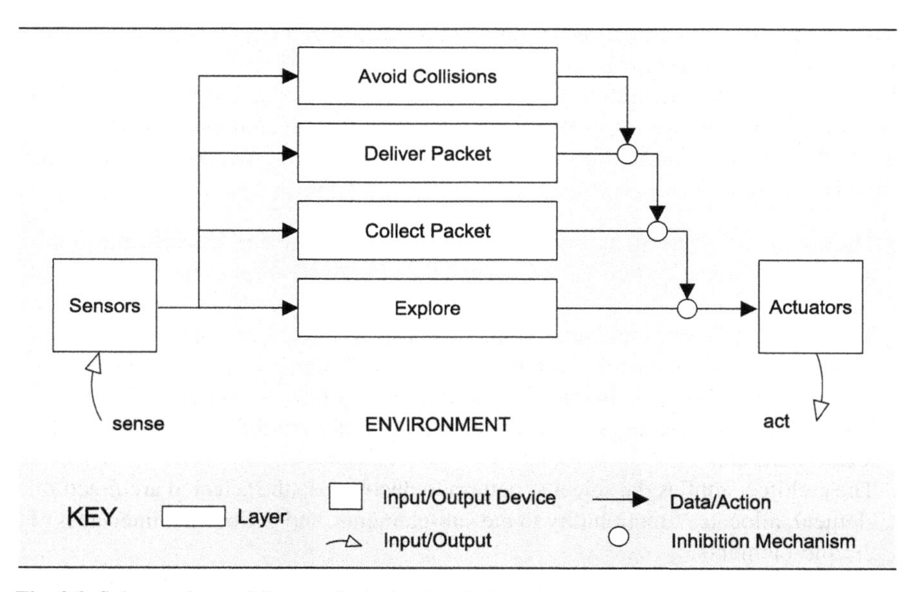

Fig. 2.3 Subsumption architecture for a simple robot

[1] Both architectural style and architectural pattern refer to recurring solutions that solve problems at the architectural design and are often used as alternatives in literature. Yet, an architectural style is looked upon in terms of components, connectors, and issues related to control and data flow. Avgeriou and colleagues argue for more attention to clarify this issue [16].

example of a subsumption architecture for a simple robot that has to collect packets and deliver them at a destination. On its way, the robot must avoid obstacles in the environment. A layer in the architecture directly connects perception to action by means of a finite state machine augmented with timing elements. Each layer collects its own sensor data that is written in registers. The arrival of specific data, or the expiration of a timer, can trigger a change of state in the interior finite state machine and possibly produce output commands to actuators. Inhibition mechanisms resolve conflicts between actuator commands from different layers. The subsumption architecture pattern allows the design of very efficient agents. However, subsumption architectures are very hard to build for complex agents that have to operate in complex environments. Nevertheless, the pattern has successfully been used in many practical robots.

It is important to notice that the design of a complex multi-agent systems typically requires a combination of agent-based patterns and other common architectural patterns. We illustrate the combined use of architectural patterns for the design of the case study in Chap. 4.

2.3.2 ADD Process

ADD takes as input the functional requirements, prioritized quality attribute scenarios, and design constraints. Examples of design constraints are the use of a particular framework, the integration with legacy systems, etc. The output of ADD is a software architecture for the system under development, documented using several views. We elaborate on architectural documentation below. ADD consists of the following steps:

1. The architect selects an architectural element for refinement. Usually, the architect starts from the system as a whole and then iteratively refines the architectural elements.
2. The architect determines the architectural drivers, i.e., a set of architecturally significant requirements that apply to the element being designed consisting of functional goals and quality attribute scenarios that have to be realized.
3. The architect selects an appropriate architectural pattern that satisfies the architectural drivers.
4. The architect applies the selected pattern to decompose the selected architectural element, allocates functionality to the sub-elements, and defines the interfaces of the sub-elements.
5. The architect refines the use cases, quality requirements, and constraints and allocates them to the newly created design elements.
6. The architect repeats steps 1–5 until the architectural elements are sufficiently fine-grained, and downstream design and implementation can start. At that point, the architectural drivers are satisfied and the software architecture becomes a prescriptive plan for construction of the system that enables effective satisfaction of the system's functional and quality requirements.

Besides the documentation of the architectural structures, it is important that the architect documents all relevant information that relates to the design decisions, including a design rationale for selected architectural patterns, how quality attributes have been satisfied, rejected alternatives, and a motivation why the alternative was rejected. The knowledge captured in this additional information allows stakeholders to understand the rationale for the design decisions and is a basis for architectural evaluation.

We elaborate on patterns in Chap. 3 when we describe how architectural patterns provide the means to capture well-proven domain expertise in multi-agent system engineering. ADD is discussed in detail in Chap. 4 when we zoom in on architectural design of multi-agent systems.

2.4 Middleware Support for Multi-Agent Systems

Popular frameworks such as Jade [23], and Jack [174] have a relative narrow view on middleware support for agent-based systems and basically provide infrastructure for communication or a broker infrastructure. Common middleware services such as security, persistency, and transactions are often considered minimally in multi-agent system development. Examples for platforms that provide some support for integration with common middleware services are Retsina [156] developed at Carnegie Mellon University that includes basic services for security, performance monitoring, logging, and failure monitoring, and the more recently developed Living Systems of Whitestein Technologies [171] that is integrated with JEE and provides support for data management with transactions, persistency, client access through Web services, etc.

Middleware support for multi-agent systems beyond communication services for message exchange, such as electronic institutions [54] and infrastructure for stigmergic interaction [35, 106], tends to be less mature. These middleware platforms have mainly been used in experimental settings and research labs. As a result, multi-agent system engineers in practice have to develop middleware that fits the needs of their particular domain. In Chap. 5, we elaborate on middleware support for multi-agent systems and we discuss in detail the middleware that was developed for the case study.

2.5 Documenting Software Architecture

To be effective, a software architecture must be well-organized and unambiguously communicated to the group of stakeholders. Therefore, good documentation of the software architecture is of utmost importance. The documentation must be not only general enough to be quickly understandable but also concrete enough to guide developers to construct the system. Clements et al. [45] gives three fundamental uses of architecture documentation:

1. Communication among stakeholders. The software architecture represents a common abstraction that serves as a primary vehicle for communication among stakeholders. Software architecture forms a basis for project organization; it imposes constraints on the design and implementation of the system; it is a starting point for maintenance activities; etc.
2. Software architecture serves as a basis for system analysis. The architecture must contain the necessary information to evaluate the various attributes; we elaborate on architecture evaluation in Sect. 2.6.
3. Architecture serves as a means for training. Software architecture is a useful instrument to introduce new people to the system, such as new team members, external analysts.

2.5.1 Architectural Views

It is generally accepted that a software architecture should be described by several *views* that emphasize different aspects of the architecture. Building upon the work of Parnas [122] and Perry and Wolf [127], Kruchten introduced four main *views* of software architecture [92]. Each view emphasizes specific architectural aspects that are useful to different stakeholders. The logical view gives a description of the services the system should offer to the end users; the process view captures the concurrency and synchronization aspects of the design; the physical view describes the mapping of the software onto the hardware and reflects its distribution aspects; and the development view describes the organization of the software and associates the software modules to development teams. A final additional view shows how the elements of the four views work together. Some other relevant work on views include [152, 143].

In architecture-based design of multi-agent systems, we follow the approach of "Views and Beyond" introduced by Clements and colleagues [45]. This approach is compatible with the ISO/IEC 42010 standard on systems and software engineering, recommended practice for architectural description of software-intensive systems [76]. In Views and Beyond, a *view type* defines the element types and relationship types used to describe the architecture of a software system from a particular perspective. Each view type constrains the set of elements and the relations that exist in its views. Three view types are distinguished:

1. The module view type: views in the module view type document a system's principal units of implementation.
2. The component-and-connector view type: views in the component-and-connector view type document a system's units of execution.
3. The allocation view type: views in the allocation view type document the relationships between a system's software and its development and execution environment.

An architectural style is a specialization of a view type and reflects a recurring architectural approach that exhibits specific quality attributes, independent of any particular system. For example, "layered style" is a specialization of the module view type. The layered style describes the system, or a part of the system, as a set of layers. Each layer is allowed to use the services of the layer below. "Communicating processes style" is an example of the component-and-connector view type. The communicating processes style describes concurrent units such as processes and threads and the connection between the units such as synchronization and control. A view is an instance of an architectural style that is bound to specific elements and relations in a particular system.

The documentation of a software architecture consists of the relevant views completed with additional information that applies to different views. What views should be documented depends on the goals of the documentation. A software architecture intended for initial project planning likely contains another set of architectural views as an architecture that specifies the implementation units for development teams. Different views highlight different system elements and their relationships and expose different quality attributes. Therefore, the views that expose the most important quality attribute requirements of the stakeholders should be part of the architecture documentation. Additional information of the software architecture documentation may include background information, a view template, a mapping between views that explains the relations between different views, a glossary, etc.

Architecture documentation with Views and Beyond is discussed in detail in Chap. 4 when we zoom in on documenting software architectures of multi-agent systems.

2.5.2 Architectural Description Languages

Architectural description languages (ADLs) [109] are languages that provide features for modeling software architectures. Most ADLs support the specification of components, connectors, and interfaces. First-generation ADLs such as Rapide [100] and Wright [8] were developed for specific domains. More recently, a number of modular ADLs have been developed such as Acme [61], xADL 2.0 [49], and π-ADL [118]. These languages emphasize reuse and support the development of domain-specific ADLs. Some authors prefer the Unified Modeling Language (UML) [4] as ADL. Unfortunately, UML does not offer first-class support for many architectural concepts such as connectors, layers, views, and view relations.

Documenting typical multi-agent system concerns such as interaction protocols, roles, and organizations sometimes requires dedicated notations, probably dedicated views. Various notations are described in literature. Examples are [180] which adopts the *pi*-calculus and object-oriented Petri nets as a formal basis to model agent architectures and [56] which provides a specification in the Z language of a core model of structural and behavioral elements of BDI agents that can be used to describe the architecture of such agents. Two popular modeling languages for multi-agent systems are Agent UML [22] (AUML) and Agent Modeling

Language [44] (AML). AUML is an extension of the Unified Modeling language (UML) that includes support for modeling protocols for multi-agent interaction, agent roles, extended UML message semantics. AML is a visual modeling language for specifying, modeling, and documenting multi-agent systems developed by Whitestein Technologies. AML is based on the UML 2.0 Superstructure, augmenting it with several new modeling concepts appropriate for capturing the typical features of multi-agent systems. Although both AUML and AML provide dedicated notations for describing multi-agent systems, they suffer from the same problems as UML, i.e., lacking first-class support for various common architectural elements.

Since no ADL provides the facilities to completely document the various view types we use to document software architectures, we employ a hybrid description language that uses UML constructs where possible. Each diagram is provided with a key that explains the meaning of the symbols used.

2.6 Evaluating Software Architecture

A software architecture is the foundation of a software system; it represents a system's earliest set of design decisions [21]. These early decisions are the most difficult to get correct, the hardest to change later, and they have the most far-reaching effects. Software architecture not only structures the system's software, but also structures the project in terms of team structure and work schedules. Due to its large impact on the development of the system, it is important to evaluate the architecture as early as possible. Modifications in initial stages of the design are cheap and easy to carry out. Deferring evaluation might require expensive changes or even result in a system of inferior quality.

Architectural evaluation is examining a software architecture to determine whether it satisfies system requirements, in particular the quality attribute requirements [46, 7, 115]. Such evaluation focuses on the most important attributes, i.e., the attributes that are most important for the system's stakeholders and those that have the largest impact on the software architecture. Architectural evaluation typically takes place when the architecture has been specified, before implementation. Experiences with a prototype implementation are invaluable for the evaluation of a software architecture. Early evaluation allows to add missing pieces, to correct inferior decisions, or to detail vaguely specified parts of the architecture, before the cost of such corrections would be too high.

In architecture-based design of multi-agent systems, we use the Architecture Tradeoff Analysis Method (ATAM) [46]. ATAM is one of the most mature approaches for software architecture evaluation currently available. ATAM is a social process aiming to achieve agreement among stakeholders. The goal of ATAM is to determine the tradeoffs and risks with respect to satisfying important quality attribute requirements. ATAM is an evaluation method that uses (1) stakeholders to determine important quality attribute requirements; (2) the architect to focus on important portions of the architecture; and (3) architectural approaches to determine potential problems. There are two groups of people involved in ATAM: the evaluation team and the stakeholders. The evaluation team conducts the evaluation and performs the analysis. The stakeholders are the people that have a particular interest in the software architecture under evaluation, such as the project manager, the architect, developers, customers, (representatives of) end users. An ATAM evaluation produces the following results:

- A prioritized list of quality attribute requirements in the form of a quality attribute utility tree.
- A mapping of architectural approaches to quality attributes. The analysis of the architecture exposes how the architecture achieves—or fails to achieve—the important quality attribute requirements.
- Risks and non-risks. Risks are potentially problematic architectural decisions, non-risks are good architectural decisions.
- Sensitivity points and tradeoff points. A sensitivity point is an architectural decision that is critical for achieving a particular quality attribute. A tradeoff point is an architectural decision that affects more than one attribute, it is a sensitivity point for more than one attribute.

A crucial document in the ATAM is the quality attribute utility tree, utility tree for short. This document is a prioritized list of quality attribute goals, formulated as scenarios. A utility tree expresses what the most important quality goals of the system are. An excerpt of the utility tree of the case study is shown in Fig. 2.4.

The root node of the tree is *utility*, expressing the overall quality of the system. High-level quality attributes form the second level of the tree. Each quality attribute

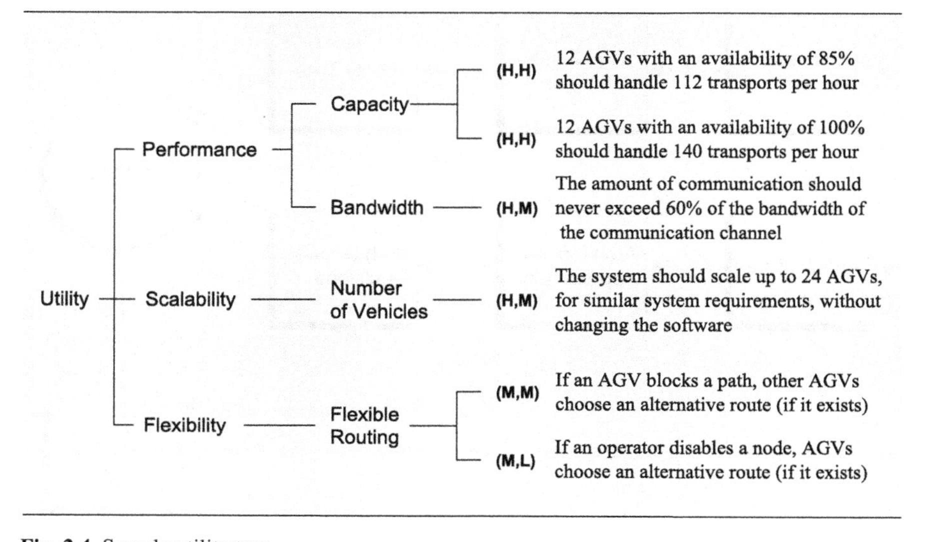

Fig. 2.4 Sample utility tree

is further refined in the third level. Finally, the leaf nodes of the tree are the quality attribute scenarios. Each scenario is assigned a ranking that expresses its priority relatively to the other scenarios, H stands for High, M for Medium, and L for Low. Prioritizing takes place in two dimensions. The first mark of each tuple refers to the importance of the scenario to the success of the system and the second mark refers to the difficulty to achieve the scenario. For example, the scenario "If an operator disables a node, AGVs choose an alternative route (if it exists)" has priorities (M,L), meaning that this scenario is of medium importance to the success of the system and relatively easy to achieve. The utility tree expresses what the most important qualities of the system are and as such it serves as a guidance for the evaluators to look for architectural approaches that satisfy the important scenarios of the system. It is clear that scenarios with priorities (H,H) and (H,M) are the prime candidates for analysis during the ATAM.

The evaluation of a software architecture with ATAM consists of three phases:

1. Presentations. The first phase consists of three steps: the evaluation leader starts by giving an overview of the evaluation method; next the project manager describes the business goals of the project; finally the architect gives an overview of the software architecture.
2. Investigation and analysis. The second phase also consists of three steps. First the architect identifies the architectural approaches applied in the software architecture. Next the quality attribute utility tree is generated. The system's quality attributes are elicited from the stakeholders and specified as scenarios. The list of scenarios is then prioritized. Finally, the architectural approaches that address the high-priority scenarios are analyzed, resulting in a list of risks, non-risks,

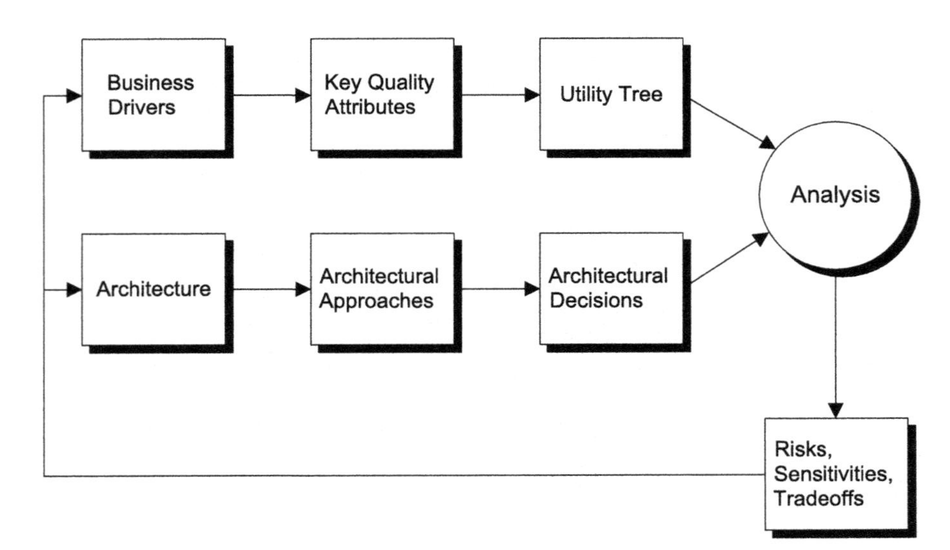

Fig. 2.5 Conceptual flow of the ATAM

sensitivity points, and tradeoff points. The analysis may uncover additional architectural approaches.

3. Reporting the results. In the final phase, the information collected during the ATAM is presented to the assembled stakeholders.

The flow of the ATAM is summarized in Fig. 2.5. The flow illustrates how the ATAM exposes architectural risks that may impact the software architecture and possibly the achievement of the organization's business goals.

The disciplined evaluation of the software architecture of a multi-agent system is invaluable in practice. It allows to clarify the qualities offered by a multi-agent system architecture. However, it also allows to pinpoint the tradeoffs with respect to other qualities and possible risks implied by adopting a multi-agent system architecture.

Architecture evaluation with ATAM is discussed in detail in Chap. 7 when we zoom in on the evaluation of software architectures of multi-agent systems.

2.7 From Software Architecture to Downstream Design and Implementation

A software architecture serves as a blueprint for system development. It defines constraints on downstream design and implementation; it describes how the implementation must be divided into elements and how these elements must interact with one another to fulfill the system goals. On the other hand, a software architecture does not *define* an implementation, many fine-grained design decisions can be left open by the architects and must be resolved by designers and developers. Examples are internal data structures of modules, specific protocols and algorithms, the use of specific object-oriented design patterns, detailed exception handling, etc.

Downstream design deals with the realization of the architectural elements (modules, components, connectors, interfaces, etc.) which are determined by the architecture. In agent-oriented software engineering, downstream design typically focuses on developing the internal capabilities of the agents, i.e., reasoning constituents, internal events, plans, and detailed data structures. Furthermore, during downstream design concrete support for agent interaction has to be developed, including a communication language and supporting services. Often, common object-oriented modeling techniques are used such as class diagrams, interaction diagrams, and statecharts. A number of agent programming languages have been developed for programming agent systems. Some languages are based on a declarative style of programming, some are based on imperative style programming, and others combine these programming styles. For example, 2APL [50] provides programming constructs to create individual agents and specify the agents' access relations to the external environment which are assumed to be implemented as Java objects. 2APL has been used to implement different auction types and negotiation mechanisms. Existing agent programming languages [59, 129, 172, 29] are primarily designed to implement agents in terms of BDI concepts (Belief, Desire, Intention). Little experiences are reported that use these languages for programming real-world applications.

By dictating how the system is divided into prescribed elements and their interactions, software architecture provides a separation of concerns. This allows management decisions to assign tasks to development teams. Each team has to conform to the specification of their individual elements allowing teams to work largely independent and interact in disciplined ways. Software architecture is a vehicle for controlled interaction among teams. It is generally acknowledged that the software architecture and the structure of the developing organization are interrelated. As a consequence, changing the software architecture typically requires corresponding changes in the way people are structured in teams for developing, testing, and maintaining the software. Facilitating the adoption of a multi-agent system architecture and investigating a suitable adoption strategy are crucial aspects of fielding a multi-agent system.

2.8 Summary

In this chapter, we gave an overview of architecture-based design of multi-agent systems. We have put architecture-based design in a software development life cycle and showed how the different methods used in the approach map on the steps of the life cycle. Then, we explained the different steps of the approach in more detail.

Quality attribute scenarios provide the means to express stakeholders' preferences about the system in a measurable manner. A utility tree provides a mechanism to define and prioritize the relevant quality requirements precisely. Quality attribute scenarios and a utility tree are the primary vehicles for the design and evaluation of the software architecture. A QAW is a facilitated method that engages stakeholders to discover and prioritize the driving quality attributes of a software-intensive system. Pinpointing the quality attributes associated with multi-agent systems and identifying possible conflicts with other quality attributes allow to clarify the added value and tradeoffs of adopting a multi-agent system.

Architectural design requires a well-founded design method. In architecture-based design of multi-agent systems, we use ADD as systematic method to design a software architecture. ADD is an iterative decomposition method that is based on understanding how to achieve quality goals through proven architectural approaches. Tactics and architectural patterns are widely used architectural approaches that have proven to be useful to achieve particular quality attributes. The design of a multi-agent system typically requires a combination of patterns for multi-agent systems and other common architectural patterns.

Well-organized architecture documentation is crucial to communicate a system's software architecture to the varied group of stakeholders. In Views and Beyond, software architectures are documented by means of different views that emphasize different aspects of the architecture. Architectural description languages provide features for modeling software architectures. Documenting multi-agent system-specific concerns such as interaction protocols, roles, and organizations may require dedicated notations or even specific views.

Due to its large impact on the development of the system, a software architecture should be evaluated as early as possible. In architecture-based design of multi-agent systems, we use ATAM to evaluate the software architecture of agent-based systems. During ATAM, the stakeholders determine the tradeoffs and risks with respect to satisfying important quality attribute requirements of the software architecture. A disciplined evaluation of software architecture allows to pinpoint the advantages as well as the tradeoffs implied by adopting a multi-agent system architecture.

By constraining downstream design and implementation, software architecture provides the foundation for allocating work to development teams and ultimately achieving the system goals. Downstream design and implementation concerns the realization of the architectural elements which are determined by the architecture. Particular aspects of downstream design of multi-agent systems include the realization of the internal architecture of agents and the mechanisms of interaction.

Chapter 3
Capturing Expertise in Multi-Agent System Engineering with Architectural Patterns

An architectural pattern is a key concept in architectural design. It specifies a generic solution scheme for a recurring design problem. A solution scheme describes a set of components, their responsibilities and relationships, and the way in which they collaborate. Architectural patterns exhibit various properties: patterns address different quality requirements, they help to document the architectural design decisions, and facilitate communication between stakeholders through a common vocabulary. A coherent set of related architectural patterns that describe good design practices within a particular domain makes up a *pattern language* [37]. Making explicit the relationships among the patterns gives the architect guidance about how to combine the patterns to construct a software architecture for a concrete system.

In this chapter, we show how architectural patterns provide the means to capture well-proven domain expertise in multi-agent system engineering. In particular, we describe a pattern language for the domain of situated multi-agent systems. Situated multi-agent systems are one family of multi-agent systems. The focus of situated agency is on direct coupling of perception to action, modularization of behavior, and dynamic interaction with the environment. This contrasts with deliberative approaches of multi-agent systems that emphasize knowledge representation and rational choice [134, 176]. The patterns for situated multi-agent systems distill and provide a means to reuse the design knowledge derived from extensive experiences with developing various multi-agent systems. The pattern language consists of five patterns: situated agent, virtual environment, selective perception, roles and situated commitments, and protocol-based communication. In Chap. 4, we explain how we have used the architectural patterns for situated multi-agent systems during architectural design in the case study.

We start this chapter by introducing situated multi-agent systems providing some background of the pattern language for them. Next, we explain the characteristics and requirements of the target domain of the pattern language. Then, we present the pattern language. We give a general overview of the patterns and their relationships, and we zoom in on the individual patterns. We conclude with a summary of the chapter.

D. Weyns, *Architecture-Based Design of Multi-Agent Systems*,
DOI 10.1007/978-3-642-01064-4_3,

3.1 Situated Multi-Agent Systems

To provide the necessary background on the pattern language for situated multi-agent systems, we give a brief sketch of the history of situated agency. We start with the early single-agent systems. Then, we explain stigmergic multi-agent systems and situated multi-agent systems.

3.1.1 Single-Agent Systems

In the mid-1980s, researchers were faced with the problem of how to build autonomous robots that are able to generate robust behavior in the face of uncertain sensors and an unpredicted environment [33]. Attempts to build such robots with traditional techniques from artificial intelligence showed deficiencies such as brittleness, inflexibility, and no real-time reaction [102, 131]. This brought a number of researchers to the conclusion that reasoning on symbolic internal models, and planning the sequence of actions to achieve the goals, is unfeasible for agents with many, often conflicting goals that have to operate in complex, dynamic environments. This conclusion led to the development of a radically new approach to build autonomous agents. A key characteristic of this approach, described by Brooks [33], is situatedness, i.e., the robots are situated in the world, they do not deal with abstract descriptions but are directly coupled with the dynamic environment which influences the behavior of the system.

The archetype architecture of reactive agents is the subsumption architecture [32]. We explained the subsumption architecture in Sect. 2.3.1. Other representative examples of approaches for reactive agents are Pengi [5] and Situated Automata [141]. In Pengi, a penguin's situated actions are coded in the form of simple rules. To formulate these rules, Pengi does not associate symbols with individual objects in the world, but uses expressions that describe causal relationships between the agent and the entities in the world. An example of a situated action is "if there is an ice-cube-besides-me then push ice-cube-besides-me." In Situated Automata, an agent program is generated from a declarative specification [80]. This program achieves real-time performance; it acts reactively without doing any symbol manipulation.

In [102], Maes points out that for complex agents in complex environments, reactive architectures are very hard to build. The designer must anticipate what the best action is to take in all occurring situations. For complex systems much of the necessary information will only be available at runtime. Goals may vary over time and new goals may come into play. Different approaches that support runtime decision making have been developed, usually referred to as behavior-based or situated agents. Pioneering examples are Maes' Agent Network Architecture [102], Motor Schemas [11], and Free-Flow Architectures [140, 158]. We illustrate action selection with a free-flow architecture below.

From the early start of situated agent systems, there has been an ongoing discussion about the exploitation of internal world models in agent architectures. Brooks argued against the need for any kind of world model or cognitive level at all [32]. In [154], Steels states that "autonomous agents without internal models will always be severely limited." Arkin [12] agrees and states that "despite the assumptions of early work in reactive control, representational knowledge is important for robot navigation," and he demonstrates how a priori and dynamically acquired world knowledge can be exploited to increase flexibility and efficiency of reactive navigation. In [104], Malcolm and Smithers introduced hybrid architectures. A hybrid architecture combines a behavior-based subsystem with a deliberative subsystem. The deliberative subsystem permits representational knowledge to be used for planning purposes in advance of execution, while the behavior-based subsystem maintains the responsiveness, robustness, and flexibility of purely reactive systems. Today, hybrid architectures are common in the domain of robotics [13].

As an illustration of a single-agent architecture, we briefly explain action selection with a free-flow tree. We use the tree for a simple robot agent, shown in Fig. 3.1.

In short, a free-flow tree is composed of *activity nodes* (in short nodes) which receive information from internal and external stimuli in the form of *activity*. The nodes feed their activity down through the hierarchy until the activity arrives at the *action nodes* (i.e., the leaf nodes of the tree) where a winner-takes-all process decides which action is selected.

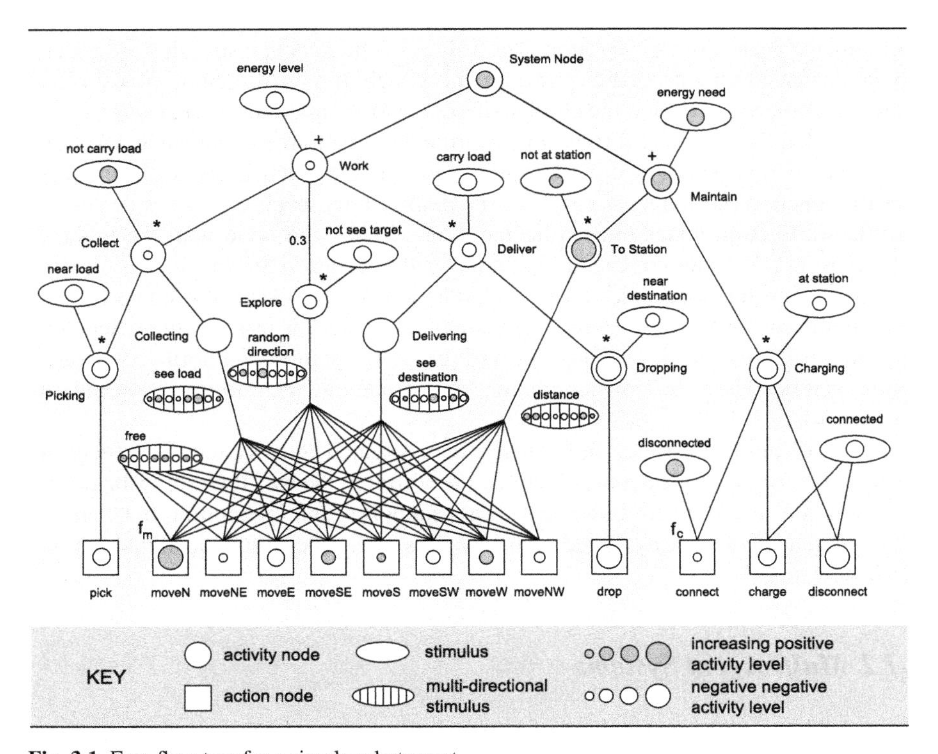

Fig. 3.1 Free-flow tree for a simple robot agent

Let us see how this works concretely. The robot we consider lives in a grid world where it has to collect loads and bring them to a destination. The robot is supplied with a battery that provides energy to work. The left part of the tree in Fig. 3.1 represents the functionality for the agent to search, collect, and deliver loads. On the right, functionality to maintain the battery is depicted. The *System Node* feeds its activity to the *Work* node and the *Maintain* node. The *Work* node combines the received activity with the activity from the *energy-level* stimulus. The "+" symbol indicates that the received activity is summed up. The negative activity of the *energy-level* stimulus indicates that little energy remains for the agent. As such the resulting activity in the *Work* node is just below zero. The *Maintain* node on the other hand combines the activity of the *System Node* with the positive activity of the *energy need* stimulus, resulting in a strong positive activity. This activity is passed to the *To Station* and the *Charging* nodes. The *To Station* node combines the received activity with the activity level of the *not at station* stimulus (the "⋆" symbol indicates they are multiplied). In a similar way the *Charging* node combines the received activity with the activity level of the *at station* stimulus. This latter is a binary stimulus, i.e., when the agent is at the charge station its value is positive (true), otherwise it is negative (false). The *To Station* node feeds its positive activity toward the action nodes it is connected with. Each moving direction receives an amount of activity proportional to the value of the *distance* stimulus for that particular direction: *distance* is a multi-directional stimulus, i.e., a compound stimulus with a value for the stimulus for each moving direction. The values of the *distance* stimulus are based on the distance to the nearest charge station for each moving direction. In a similar way, the *Charging* node and the child nodes of the *Work* node (*Explore*, *Collect*, and *Deliver*) feed their activity further downward in the tree to the action nodes. Action nodes that receive activity from different nodes combine that activity according to a specific function (f_m and f_c) to calculate the final activity level.

When all action nodes have collected their activity, the node with the highest activity level is selected for execution. In the example, the *To Station* node is clearly dominant over the other nodes connected to actions nodes. Currently the northeast, east, southwest, and northwest directions are blocked (see the *free* stimulus), leaving the agent four possibilities to move toward the charge station: via north, southeast, south, or west. The values of the gradient field guide the agent to move northward see Fig. 3.1.

As this example illustrates, initial research on situated agent systems was focused on architectures of single agents. Architectures differ in the way they solve the problem of action selection. Architectures do not support social interaction. In Chap. 4, we explain how we have extended free-flow trees with support for social interaction in the case study.

3.1.2 Multi-Agent Systems

From the late 1980s, researchers of situated agents have been investigating systems in which multiple agents work together to realize the system's goals. In these systems, infrastructure for indirect coordination has a central role. The coordination infrastructure enables agents to share information and coordinate their behavior.

In [123], Parunak describes how principles of different natural agent systems (ants, wasps, wolves, etc.) can be applied to build artificial agent systems. The underlying principle is called stigmergy, a concept introduced by Grassé [65]: one individual modifies the environment and others respond to the modification, and modify it in turn. A classic example of stigmergic coordination in agent systems is a digital pheromone infrastructure [35, 27]. A digital pheromone is a dynamic structure that is situated in a virtual environment. The pheromone aggregates with additional pheromone that is dropped, diffuses in space, and evaporates over time. Agents can use pheromones to dynamically form paths to locations of interest. Another well-established approach of stigmergic coordination is computational fields [106]. In this approach, the movements of agents are driven by abstract force fields that are spread in a virtual environment. Agents coordinate their behavior by following the shape of the fields. Dynamics in the external world and movements of the agents induce changes in the surface of the fields, realizing a feedback cycle that influences the agents' behavior.

Although stigmergic agent systems have proven their value in practice, a number of comments are in order: (1) stigmergic agents are considered as "simple" entities. However, there is little or no attention for the architecture of agents; (2) stigmergic agents are not able to set up explicit collaborations to exploit contextual opportunities; (3) infrastructures for stigmergic coordination provide reusable solutions that can be applied over many applications. Yet, choosing for a particular infrastructure compels an engineer to a specific form of coordination which may restrict flexibility.

Motivated by these considerations, researchers have extended the vision of stigmergic agents and developed architectures for a family of agent systems that is commonly referred to as situated multi-agent systems. We briefly discuss two representative approaches.

In [58], Ferber and Müller propose a model for situated multi-agent systems that builds upon earlier work of Genesereth and Nilson [63]. Ferber and Müller distinguish between tropistic and hysteric agents. Tropistic agents are essentially reactive agents without memory, whereas hysteric agents may have complex behaviors that use past experiences for decision making. Central in the approach is a model for action. This model distinguishes between influences and reactions to influences. Influences are produced by agents and are attempts to modify the course of events in the world. Reactions, which result in state changes, are produced by the environment by combining influences of all agents, given the state of the environment and the laws of the world. In [57], Ferber uses the BRIC formalism (Block-like Representation of Interactive Components) to model situated multi-agent systems based on the model for situated multi-agent systems. In BRIC, a multi-agent system is modeled as a set of interconnected components that can exchange messages via links. BRIC components encapsulate their own behavior and can be composed hierarchically.

Multilayered multi-agent situated system [18] (MMASS) defines agent types and an explicit model of the environment. The definition of an agent type comprises agent state, perceptual capabilities, and a behavior specification. Agent behavior can be specified with a behavior specification language [17] that defines a number of basic primitives, such as transport (defines a movement of the agent) and trigger (specifies state change when a particular condition is sensed in the environment). In MMASS, the environment is explicitly modeled as a multilayered structure, where each layer is represented as a connected graph of sites (a site is a node of the graph in a layer of the environment). Layers may represent abstractions of a physical environment, but can also represent logical aspects, e.g., the organizational structure of a company. Between the layers specific connections (interfaces) can be defined that are used to specify that information generated in one layer may propagate into other layers. In MMASS, agents can (1) interact through a reaction with agents in adjacent sites (a reaction is a synchronous change of state of the involved agents), (2) emit fields that are diffused in the environment, (3) perceive other agents, (4) update their state, and (5) move to adjacent sites. MMASS has been applied in various application areas, examples are an adaptive web application [28] and a distributed collaboration system [99].

3.2 Target Domain of the Pattern Language for Situated Multi-Agent Systems

The brief historical overview shows that situated agent systems have been studied and built for over two decades. The pattern language for situated multi-agent systems builds upon this foundation and integrates our experiences with the design of practical situated multi-agent systems.

The objective of the pattern language is to document well-proven design expertise and reuse this knowledge to support architectural design of situated multi-agent systems. The pattern language for situated multi-agent systems embodies the expertise we gained with the architectural design of various practical applications. We extensively used the Packet-World, a simple robotic application, as a study case for investigation and experimentation [165, 166, 168]. We derived expertise from the design and development of a peer-to-peer file sharing system [147, 170]. This application applies a pheromone-based approach for the coordination of agents that move around in a dynamic network searching for files. In [164] we have applied a field-based approach for adaptive task assignment in a mobile environment, and in [153, 169] we have applied situated multi-agent systems in several experimental robotic applications. Finally, [66, 67] use situated agents in an intelligent transportation system for monitoring traffic jams. In the course of building the various applications, we derived common functions and structures that provided architectural building blocks for the patterns of the pattern language.

The key characteristics and requirements shared by the family of software systems supported by the pattern language for situated multi-agent systems are

- Important stakeholder requirements are flexibility (adapt to variable operating conditions) and openness (cope with parts that come and go during execution).

These quality requirements may conflict with other important stakeholder requirements.
- The software systems operate under highly dynamic and changing operating conditions, such as dynamically changing workloads and variations in availability of resources and services. An important requirement of the software systems is to manage the dynamic and changing operating conditions autonomously.
- Global control is hard to achieve. Activity in the systems is inherently localized, i.e., global access to resources is difficult to achieve or even infeasible. The software systems are required to deal with the inherent locality of activity.

Typical example domains are mobile and ad hoc networks, automated transportation systems, and robotics.

3.3 Overview of the Pattern Language

Figure 3.2 shows a general overview of the pattern language for situated multi-agent systems with the most important relationships between the proposed patterns.

The basic patterns of the pattern language are *situated agent* and *virtual environment*. A situated agent is an autonomous problem-solving entity in the system. An agent encapsulates its state and controls its behavior. The responsibility of an agent is to achieve its design objectives, i.e., to realize the application-specific goals it is assigned. Agents are able to adapt their behavior according to the changing circumstances in the environment. A situated agent is a cooperative entity. The overall application goals result from interaction among agents, rather than from sophisticated capabilities of individual agents. Agents are situated in a virtual environment. The virtual environment maintains a virtualization of the relevant parts of the world and serves as a coordination medium for the agents, i.e., the virtual environment mediates both the interactions among agents and the access to resources.

Selective Perception enables a situated agent to sense its neighborhood and update its knowledge about the world. *Protocol-Based Communication* enables situated agents to exchange messages according to prescribed communication protocols, i.e., well-defined sequences of messages. *Roles and Situated Commitments* are social attitudes of situated agents. A role represents a coherent part of functionality of a situated agent in the context of an organization. A situated commitment defines a relationship between roles, providing the means to establish collaborations among situated agents. A situated commitment affects the behavior of the agents involved in the commitment in favor of the roles the agents play in the commitment.

Some of the patterns in the pattern language are optional. For example, for the design of agents that do not communicate by exchanging messages, the Protocol-Based Communication pattern can be omitted. We elaborate on a number of options in the pattern language when we present the various patterns.

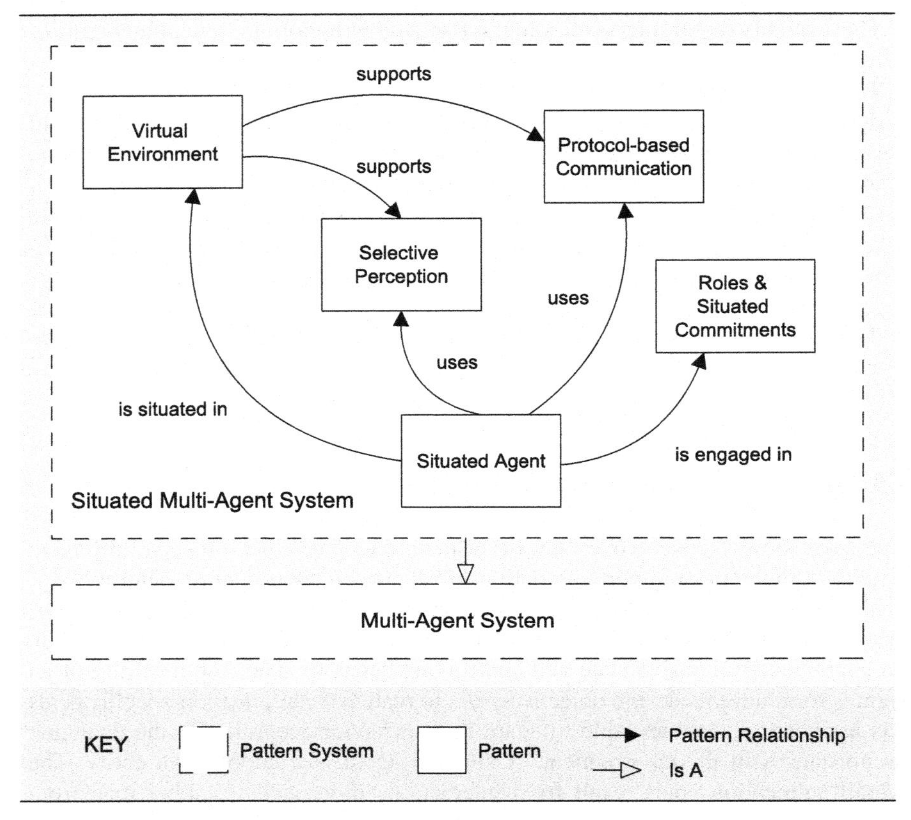

Fig. 3.2 Map of the pattern language

3.4 Pattern Template

Before we explain the patterns in detail, we first describe the organization that the documentation of each pattern obeys. A pattern of the pattern language consists of the following parts:

1. The name of the pattern.
2. A primary presentation that shows the elements and their relationships in the pattern. We use component and connector models to describe the patterns' units of execution.
3. A description of the architectural elements with their specific properties.
4. Interface descriptions that specify how the elements are used with one another.
5. An architecture rationale that explains the motivation for the design of the pattern.

The pattern template we use is inspired by the approach for documenting architectural styles presented in [45].

Appendix A gives a rigorous specification of the elements and how they are used with one another for the two basic patterns: virtual environment and situated agent.

3.5 Virtual Environment

3.5.1 Primary Presentation

The primary presentation of the virtual environment pattern is shown in Fig. 3.3. Virtual environment comprises a single data repository: State and five components—Synchronization, Dynamics, Perception Service, Action Service, and Communication Service.

3.5.2 Architectural Elements

The *State* repository has a central role in the virtual environment. The repository contains data that is shared between the components of the virtual environment. Data stored in the state repository typically includes an abstract representation of external resources and additional state related to the virtual environment. Examples of state related to external resources are a representation of the local topology of a network and data derived from a set of sensors. Examples of additional state are the representation of digital pheromones that are deployed on top of a network and virtual marks situated on the map of the physical environment. The state repository may also include agent-specific data, such as the agents' identities, the positions of the agents, and tags used for coordination purposes.

Synchronization is responsible for synchronizing state of the virtual environment with state of particular external resources as well as state of the virtual environments on neighboring nodes. An example of the former is the topology of a dynamic network in which changes are reflected in a network abstraction maintained in the state of the virtual environment. An example of the latter is the maintenance of a list of available resources that are shared among neighboring nodes. The synchronization component may pre-process the collected information before it updates the state of the virtual environment. A typical way to collect data in a distributed setting is by using a suitable middleware. In Chap. 5 we discuss a middleware that we used in the case study. This middleware supports the management of data collection in a mobile setting.

Dynamics is responsible for maintaining processes in the virtual environment that happen independent of agents and external resources. The dynamics component directly accesses the state of the virtual environment and maintains the state according to its application-specific definition. A typical example is the maintenance process of digital pheromones. State changes resulting from updates by the dynamics component may trigger the synchronization component to update the state of the virtual environment on other nodes.

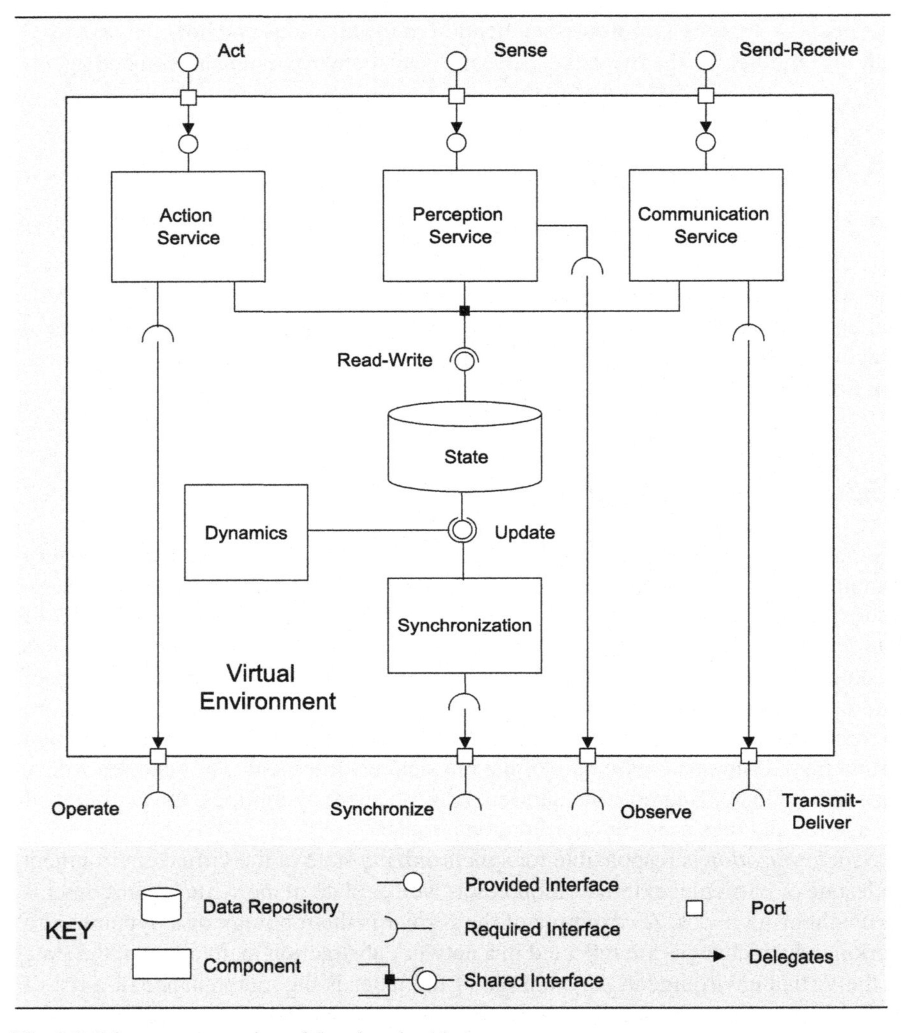

Fig. 3.3 Primary presentation of the virtual environment pattern

The *Perception service* provides the functionality to agents for sensing their neighborhood, resulting in a representation. A representation is a data structure that represents the sensed elements in a form that can be interpreted by an agent. The perception service supports *selective perception*. Selective perception enables an agent to direct its perception at the relevant aspects according to its current task. This facilitates better situation awareness and helps to keep processing of perceived data under control. To direct its perception an agent selects a set of *foci*. Each focus of the set of selected foci is characterized by a particular perceptibility, but may have other characteristics too, such as an operating range and a resolution. Examples for a robot agent are a focus to observe nearby robots and a focus to locate the charging stations within a certain distance. Focus selection enables an agent to direct its perception for specific types of information. When an agent invokes a sense request, the perception service collects the required information from the state repository of the virtual environment or from external resources (via the Observe interface). A request from the state repository may trigger the synchronization component to update the state of the virtual environment with the state of the virtual environment on neighboring nodes. The perception service can provide additional functions to pre-process raw data retrieved from external resources, such as sorting and integrating sensor data.

Action service is responsible for dealing with agents' actions. Actions can be divided into two classes: actions that attempt to modify state of the virtual environment and actions that attempt to modify the state of external resources. An example of the former is an agent that drops a digital pheromone in a pheromone infrastructure that is maintained by the state repository of the virtual environment. An example of the latter is an agent that writes data in an external database. An action that modifies the state of the virtual environment may trigger the synchronization component to update the state of the virtual environment with the state of the virtual environment on other nodes. The action service can provide additional functions to translate actions related to external resources to low-level operations.

The *Communication service* is responsible for managing the communicative interactions among agents. It is responsible for collecting messages, it provides the necessary infrastructure to buffer messages, and it delivers messages to the appropriate agents. An agent communication message typically consists of a header with the message performative (inform, request, propose, etc.), followed by the subject of this performative, i.e., the content of the message that is described in a content language that is based on a shared ontology. Such message descriptions enable a designer to express the communicative interactions between agents independent of the applied communication technology. However, to actually transmit the messages, the communication service makes use of a distributed communication system provided by an underlying middleware or communication framework. The communication service translates message descriptions used by agents to communication primitives of the supporting communication system and vice versa. Depending on the application requirements, the communication service may provide specific communication services to enable the exchange of messages in a distributed setting, such as white and yellow page services.

3.5.3 Interface Descriptions

The interface descriptions specify how the components of the virtual environment are used with one another.

The state repository exposes two interfaces. The provided interface `Update` enables attached components to read state of the repository. The `Read-Write` interface enables the attached components to access and modify the virtual environment's state.

The `Sense` interface of the virtual environment provides an operation that enables agents to perform a perception request based on a set of foci. Perception requests are delegated to the perception service. To collect data, the perception service requires the `Read-Write` interface that is provided by the state repository. To observe external resources, the perception service delegates requests to the abstract `Observe` interface that is provided by the underlying infrastructure. The concrete operations provided by the `Observe` interface are application specific.

To synchronize the state of the virtual environment with external resources, the synchronization component depends on the `Synchronize` interface that is provided by underlying infrastructure. The concrete operations provided by the `Synchronize` interface are application specific. To update the state of the virtual environment, the Synchronization component requires the `Update` interface that is provided by the state repository.

To maintain dynamics in the virtual environment that happen independent of agents or external resources, the dynamics component requires the `Update` interface that is provided by the state repository.

The `Send-Receive` interface of the virtual environment provides operations to an agent for exchanging messages with other agents. The virtual environment delegates communication requests to the communication service. The communication service requires the `Read-Write` interface of the state repository to collect data for converting and sending the messages to the addressees. The communication service delegates the transmission of messages to the `Transmit-Deliver` interface that is provided by the underlying infrastructure. The concrete operations provided by the `Transmit-Deliver` interface are application specific. The `Transmit-Deliver` interface passes incoming messages to the communication service which delivers the messages to the addressees via the `Send-Receive` interface.

The `Act` interface of the virtual environment provides operations for agents to invoke actions. The `Act` interface delegates actions to the action service. Actions that attempt to modify the state of external resources are delegated by the action service to the abstract `Operate` interface that is provided by the underlying infrastructure. The concrete operations provided by the `Operate` interface are application specific.

3.5.4 Design Rationale

The two primary concerns that underlie the design of the virtual environment pattern are use of a shared data style to decouple the various components and decomposition of functionality driven by the principle of separation of concerns.

The shared data style results in low coupling among the components, improving modifiability (changes in one element do not affect other elements or the changes have only a local effect), and reuse (elements are not dependent on too many other elements). Low-coupled elements usually have clear and separate responsibilities, which makes the elements better to understand in isolation. Decoupled elements do not require detailed knowledge about the internal structures and operations of the other elements. Due to the concurrent access of the state repository, the shared data style requires special efforts to synchronize data access.

Action service, perception service, and communication service provide operations corresponding to the various ways situated agents can access the virtual environment. The services use data local to the virtual environment and access external resources via the underlying infrastructure. Synchronization is responsible for synchronizing the state of the virtual environment with external resources and dynamics is responsible for activities private to the virtual environment. By separating the various concerns, the decomposition of the virtual environment yields a flexible modularization that can be tailored to a broad family of application domains. For instance, for applications in which agents interact via marks in the virtual environment but do not communicate via message exchange, the communication service can be omitted. For applications in which there are no dynamic processes, the dynamics component can be omitted. Minimizing the overlap of functionality among modules helps the architect to focus on one particular aspect of the functionality of the virtual environment. It supports reuse, and it further helps to accommodate change and to update one component without affecting the others.

3.6 Situated Agent

3.6.1 Primary Presentation

The primary presentation of the situated agent pattern is shown in Fig. 3.4. Situated agent comprises a single data repository: Current Knowledge and three components: Perception, Decision Making, and Communication.

3.6.2 Architectural Elements

The *Current Knowledge* repository contains state that is shared among the data accessors: Perception, Decision Making, and Communication. We distinguish between shared state and internal state. Both kinds of state can be further divided into static state and dynamic state.

- *Shared state* refers to state that is shared with other agents. Static shared state refers to the agent's state of the system that does not change over time. A typical example is a map of the environment. Dynamic state relates to state about the agent's current context; it dynamically changes over time. Examples are locally perceived objects in the environment and data about a temporal agreement for collaboration that is exchanged via messages.
- *Internal state* refers to the agent's state that is not shared with other agents. Internal state can be static or it can dynamically change over time. Examples of internal static state are the various parameters of a behavior-based action selection mechanism. An example of internal state that dynamically changes is state that

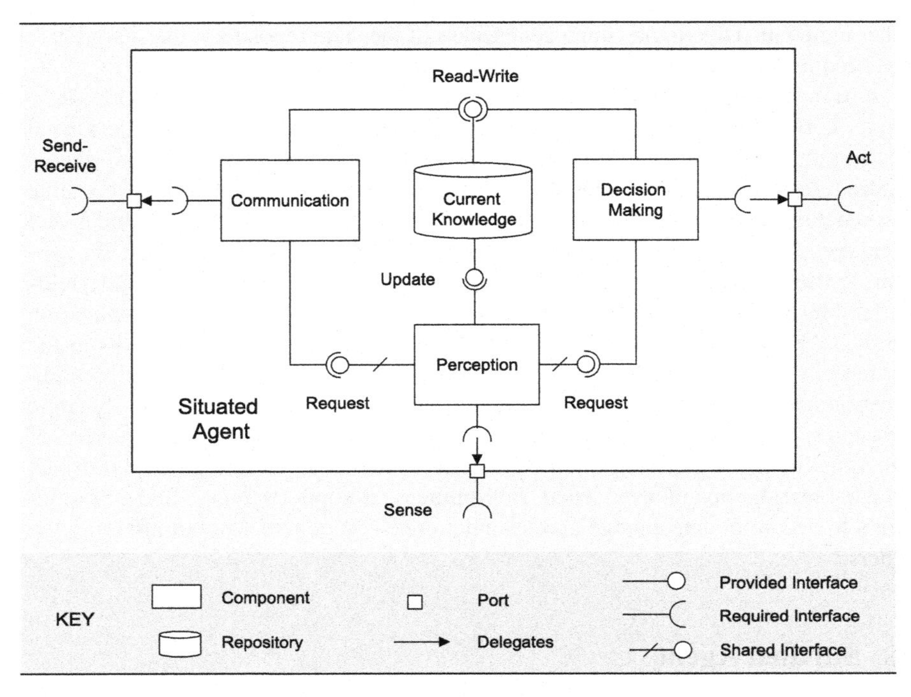

Fig. 3.4 Primary presentation of the situated agent pattern

represents the success rate of recently selected behaviors. An agent can use such state to adapt its behavior over time, see, e.g., [168].

Perception is responsible for collecting runtime information from the virtual environment. The perception component supports selective perception, enabling an agent to direct its perception to its current tasks. Perception requests are triggered by the communication component or the decision making component. A perception request includes a set of selected *foci* and a set of selected *filters*. The perception component uses the foci to sense the virtual environment for specific types of information. Sensing results in a representation of the sensed environment. A representation is a data structure that represents elements in the virtual environment or external resources. The perception component interprets the representation resulting in a percept. A percept consists of data elements that can be used to update the agent's current knowledge. The selected set of filters reduces the percept according to the criteria specified by the filters before it updates the current knowledge. We elaborate on perception when we discuss the selective perception pattern.

The *Decision making* component encapsulates a behavior-based action selection mechanism. Decision making is responsible for realizing the agent's tasks by invoking actions in the virtual environment. To enable situated agents to set up collaborations, behavior-based action selection mechanisms are extended with the notions of role and situated commitment. We elaborate on behavior-based action selection when we discuss the roles and situated commitments pattern.

Communication is responsible for communicative interactions with other agents. Message exchange enables agents to share information directly and set up collaborations. The communication module processes incoming messages and produces outgoing messages according to well-defined communication protocols. A communication protocol specifies a set of possible sequences of messages. The information exchanged via a message is encoded according to a shared communication language. The communication language defines the format of the messages, i.e., the subsequent fields the message is composed of. Communicative interactions among agents are based on an ontology that defines a shared vocabulary of words that agents use in messages. The ontology enables agents to refer unambiguously to concepts and relationships between concepts in the domain when exchanging messages. We elaborate on communication when we explain the protocol-based communication pattern.

3.6.3 Interface Descriptions

The interface descriptions specify how the components of a situated agent are used with one another.

The current knowledge repository exposes two interfaces. The provided interface `Update` enables the perception component to update the agent's knowledge according to the information derived from sensing the virtual environment. The `Read-Write` interface enables the communication and decision making components to access and modify the agent's current knowledge.

The provided `Request` interface of the perception component enables decision making and communication to sense the virtual environment according to their current activities. Therefore, decision making and communication pass on a set of selected foci and a set of selected filters to the perception module.

The perception component's required `Sense` interface is delegated to the agent's required `Sense` interface. Similarly, the `Send-Receive` interface of the communication component and the `Act` interface of the decision making component are delegated to the required interfaces of agent with the same name. The ports decouple the internals of the agent subsystem from external elements.

3.6.4 Design Rationale

In a situated multi-agent system, control is divided among the agents. Situated agents manage the dynamic and changing operating conditions locally and autonomously. Both are important properties of the target applications of the pattern language. However, decentralized control and locality imply a number of tradeoffs and limitations;

- Decentralized control in distributed systems typically requires more communication. The performance of the system may be affected by the communication links between agents.

- There is a tradeoff between the performance of the system and its flexibility to handle disturbances. A system that is designed to cope with many disturbances generally needs redundancy, usually to the detriment of performance.
- Agents' decision making is based on local information only, which may lead to suboptimal system behavior.

These tradeoffs and limitations should be kept in mind throughout the design and development of a situated multi-agent system. Special attention should be payed to communication which could impose a major bottleneck.

The collaboration among the components of a situated agent contributes to the adaptability of the system.

Perception on Command. Selective perception enables an agent to focus its attention to the relevant aspects in the environment according to its current tasks. When selecting actions and communicating messages with other agents, decision making and communication typically request perceptions to update the agent's knowledge about the environment. By selecting an appropriate set of foci and filters, the agent directs its attention to the current aspects of its interest and adapts it attention when the operating conditions change.

Coordination Between Decision Making and Communication. The overall behavior of the agent is the result of the coordination of two modules: decision making and communication. Decision making is responsible for selecting suitable actions. Communication is responsible for the communicative interactions with other agents. However, the two components coordinate to complete the agent's tasks more efficiently. For example, agents can send each other messages with requests for information that enable them to act more purposefully. Decision making and communication also coordinate during the progress of a collaboration. Collaborations are typically established via message exchange. Once a collaboration is achieved, the communication module activates a situated commitment. This commitment will affect the agent's decision making toward actions in the agent's role in the collaboration. This continues until the commitment is deactivated and the collaboration ends. We elaborate on situated commitments below.

Ensuring that both decision making and communication behave in a coordinated way requires a careful design. On the other hand, the separation of functionality for coordination (via communication) from the functionality to perform actions to complete tasks has several advantages, as listed above (clear design, improved modifiability, and reusability). Two particular advantages of separating communication from performing actions are as follows: (1) it allows both functions to act in parallel and (2) it allows both functions to act at a different pace. In many applications, sending messages and executing actions happen at different tempo. A typical example domain is robotics, but it applies to any application in which the time required for performing actions in the environment differs significantly from the time to communicate messages. Separation of communication from performing actions enables agents to reconsider the coordination of their behavior while they perform actions, improving adaptability and efficiency.

3.7 Selective Perception

3.7.1 Primary Presentation

The primary presentation of the selective perception pattern is shown in Fig. 3.5. Selective perception comprises one data repository: Descriptions and three components: Sensing, Interpreting, and Filtering.

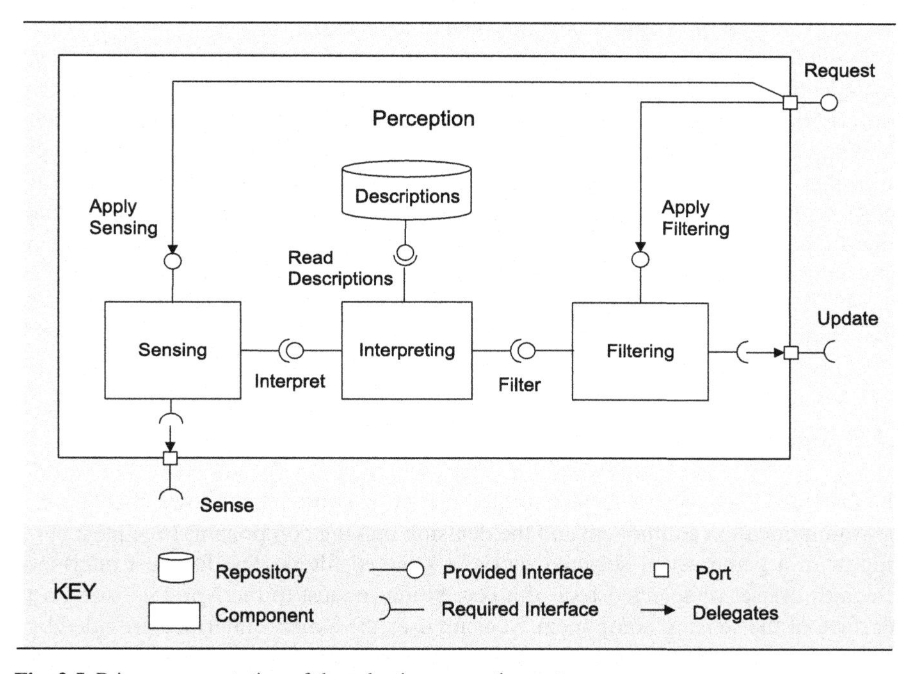

Fig. 3.5 Primary presentation of the selective perception pattern

3.7.2 Architectural Elements

To explain the collaborations between the various elements, we follow the logical thread of successive activities that take place from the moment an agent takes the initiative to sense the virtual environment until the percept is available to update the agent's current knowledge.

Sensing takes a set of foci to produce a perception request that is passed to the virtual environment. As a result, the virtual environment produces a representation. We have explained the concepts of a focus and a representation in the discussion of the virtual environment pattern.

The *Descriptions* repository contains a set of *descriptions* to interpret the given representation. A description provides a template that specifies a particular pattern of a representation. Consider for example a representation that represents a number of objects in a certain area. When the interpreting component interprets this representation it may use one description to interpret the distinguished objects and another description to interpret the group of objects as a cluster.

The *Interpreting* component uses the descriptions to extract a *percept* from the representation. A percept consists of data elements that describe elements sensed in the virtual environment or external resources in a form that can be used to update the current knowledge of the agent. Each match between the description template and the examined representation yields data of a percept.

The *Filtering* component filters a percept using set of selected *filters*. Filters allow the agent to select only those data elements of a percept that match specific selection criteria. Each filter imposes conditions on a percept that determine whether the data elements of the percept can pass the filter or not. For example, a robot agent that has selected a focus to perceive objects in its environment and that is only interested in the location of a particular type of objects can select a filter that selects the data elements with the locations of that type of objects—at least, if such data element was part of the original percept, otherwise the resulting percept will be empty. The filtering component uses the filtered percept to update the agent's current knowledge.

3.7.3 Interface Descriptions

The provided `Request` interface of the perception component allows clients (i.e., the communication component and the decision making component) to request percepts with a given set of selected foci and selected filters. The Request interface delegates the set of selected foci of a perception request to the `ApplySensing` interface of the sensing component. Sensing uses the `Sense` interface provided by the virtual environment to invoke the perception request. Sensing passes the resulting representation to the interpreting component using the `Interpret` interface. The descriptions repository exposes the `ReadDescriptions` interface. Interpreting requires this interface to interpret representations. The resulting percept is passed to the filtering component using the `Filter` interface. Filtering uses the set of selected filters provided by the `Request` interface to filter the percept. Filtering uses the `Update` interface provided by the current knowledge repository to update the agent's knowledge with the filtered percept.

3.7.4 Design Rationale

The integrated set of components of perception provides the functionality for selective perception of a situated agent. The overall functionality results from the collaboration of the various components. In this collaboration, each component provides a clear-cut functionality, while the coupling between the components is kept low.

Foci, descriptions, and filters are considered as first-class elements in the pattern. This helps to improve modifiability and reusability. The interpreting component can be omitted in case the internal state of the agent and the observable state of the virtual environment are represented by the same data types.

3.8 Roles and Situated Commitments

3.8.1 Primary Presentation

The primary presentation of the roles and situated commitments pattern is shown in Fig. 3.6.

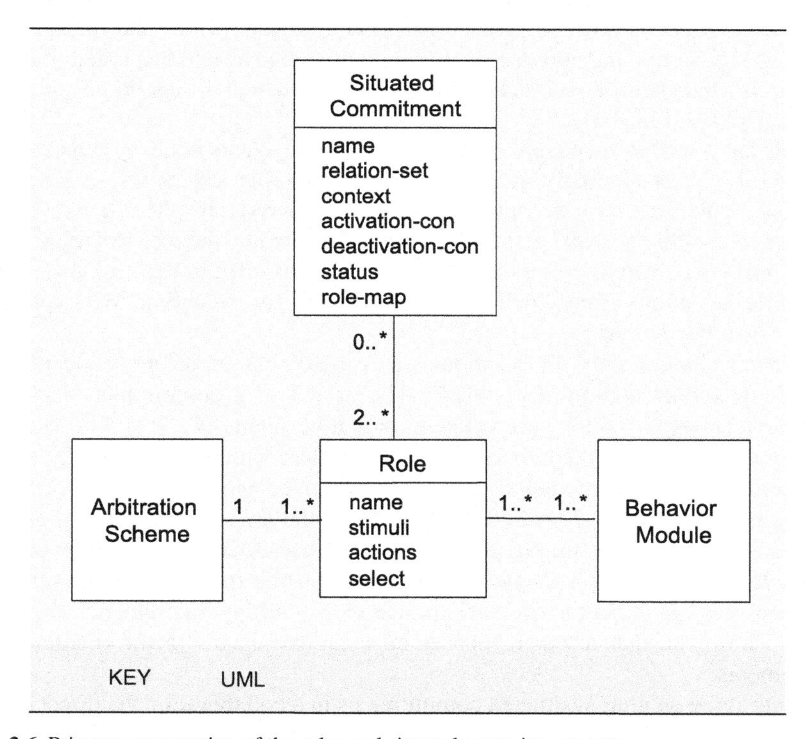

Fig. 3.6 Primary presentation of the roles and situated commitment pattern

3.8.2 Architectural Elements

To select actions, a situated agent employs a behavior-based action selection mechanism. The main advantages of behavior-based action selection mechanisms are efficiency and flexibility to deal with dynamism in the environment. In general, a behavior-based action selection mechanism consists of a set of *behavior modules*. Each behavior module is a relatively simple computation module that tightly couples sensing to action. An *arbitration scheme* controls which behavior-producing module has control and selects the next action of the agent.

As we explained in Sect. 3.1.1, behavior-based action selection mechanisms are developed from the viewpoint of individual agents. Yet, in a situated multi-agent system it is often desirable to endow agents with abilities for explicit social interaction. Explicit social interaction enables agents to exchange information directly with one another and set up collaborations. The roles and situated commitments pattern provide the means for situated agents to set up collaborations.

Role. Behavior modules that represent a coherent part of an agent's functionality in the context of an organization are denoted as a *role*. We consider an organization as a group of agents that can play one or more roles and that work together. Roles are the building blocks for social organization of a multi-agent system. This perspective on roles is similar to other approaches in agent research provided that collaborations between situated agents are bounded to the locality in which the agents are situated, see, e.g., [83, 39, 114, 107].

A role has a well-known *name* that is shared among agents in the system. *stimuli* are internal data or externally perceived information that affects the selection of actions of a role. Based on the actual stimuli, *select* determines the relative preferences for each of the possible actions that can be selected by the role. An arbitration schema uses the relative preferences for all actions of all the roles to determine which role has control and which action is selected for execution. We explain a concrete example below.

Situated Commitment. Collaborations are explicitly communicated cooperations reflected in mutual commitments [125]. The notion of a commitment has been studied extensively from the perspective of cognitive agents, see, e.g., [47, 94, 51]. Contrary to the traditional approaches on commitment which are essentially based on the mutually dependent mental states of the involved agents and a goal-oriented plan, a *situated commitment* is defined in terms of the *roles* of the involved agents and the local *context* they are placed in. Agreeing on a situated commitment incites a situated agent to give preference to the actions in the role of the commitment. This perspective is related to the sociological viewpoint on commitment proposed in [150]; however, that research focuses on cognitive agents in information-rich environments.

Agents agree on mutual situated commitments in a collaboration via direct communication (see Sect. 3.9). Once the agents have agreed on a collaboration, the mutual situated commitments will affect the selection of actions in favor of the agents' roles in the collaboration. It is important to notice that an agent can also commit to itself. For example, if a robot runs out of energy, the robot agent can commit to itself to resolve this urgent problem. Once committed, the agent will select actions in the role to recharge its battery. The commitment ends when the battery is recharged.

As for roles, situated commitments have a well-known *name*. Explicitly naming roles and commitments enables agents to set up collaborations, reflected in mutual commitments. The *relation-set* contains the identities of the related agents in the situated commitment. The *context* describes contextual properties of the situated commitment such as descriptions of objects in the local environment. *Activation-con* and *deactivation-con* are the activation and deactivation conditions that determine the *status* of the situated commitment. When the activation condition becomes true, the situated commitment is activated. The behavior of the agent will then be biased according to the specification of the *role-map*. The *role-map* specifies the relative weight of the preferences of the actions of different roles. In its simplest form, the *role-map* narrows the agent's action selection to actions in one particular role. An advanced example is a *role-map* that biases action selection toward the actions of one role relative to the preferences of the actions of a number of other roles of the agent. As soon as the deactivation condition becomes true, the situated commitment is deactivated and will no longer affect the behavior of the agent.

3.8.3 Design Rationale

Behavior-based action selection enables agents to behave according to the situation in the environment and flexibly adapt their behavior with changing circumstances. The notions of a role and situated commitment enable agents to set up collaborations. Whereas traditional approaches of commitment impose agents to communicate explicitly when the conditions for a committed cooperation no longer hold, for a situated commitment it is typically the local context in which the involved agents are placed that regulates the duration of the commitment. For example, when two robot agents form a chain to transport loads, the collaboration ends when no more loads are left to pass on or when one of the agents leaves its post for maintenance. This approach fits the general principle of situatedness in situated multi-agent systems and improves flexibility and openness. An agent adapts its behavior when the conditions in the environment change or when agents enter or leave its scope of interaction. In the next section, we illustrate how a free-flow tree, a concrete behavior-based action selection mechanism, is extended with roles and situated commitments.

3.8.4 Free-Flow Trees Extended with Roles and Situated Commitments

In Sect. 3.1.1, we explained action selection with a free-flow tree for a single robot agent. Now, we show how the roles and situated commitment pattern is used to extend free-flow trees with support for roles and situated commitments enabling explicit collaborations among multiple agents. To illustrate the explanation, we use a simple grid world in which robot agents are situated. The task of the agents is to transport loads from one location to another. To improve efficiency, robot agents can form a chain and pass loads to one another. Figure 3.7 shows a simplified free-flow tree of a robot agent.

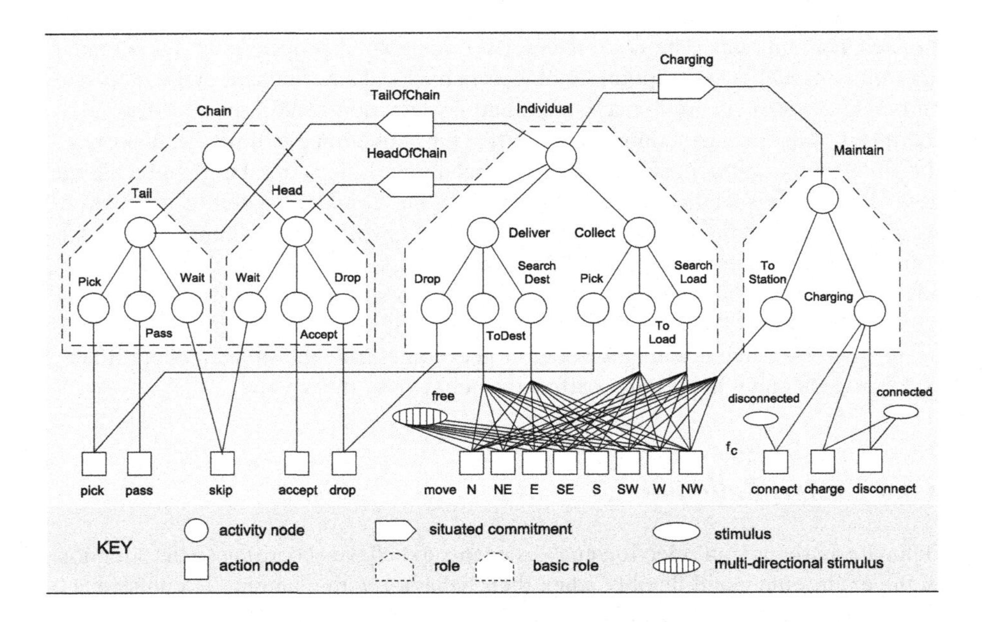

Fig. 3.7 Free-flow tree for a robot agent with roles and situated commitments (system node, combination functions, and stimuli of nodes are omitted)

A role corresponds to a subtree in the hierarchy. In the example, the roles are demarcated by dashed lines. A role is named as the root node of the subtree that represents the role. Roles that are not further divided into sub-roles are called basic roles. A robot agent has three main roles: *Individual*, *Chain*, and *Maintain*. In the role *Individual*, the agent performs work, independent of the other robot agents. The agent searches for loads and brings them to the destination. The *Chain* role is composed of two sub-roles: *Head* and *Tail* denoting the two roles of agents in a collaboration to pass loads along a chain.[1] Finally in the *Maintain* role, the agent recharges its battery. All roles of the agent are constantly active and contribute to the final decision making by feeding particular sets of actions with activity. However, the contribution of each role depends on the activity it has accumulated from the stimuli of its nodes.

A situated commitment is represented by a connector between roles in the tree. The connector *Charging* in Fig. 3.7 denotes the situated commitment of an agent to itself to recharge its battery. *Charging* connects the top nodes of the source roles *Individual* and *Chain* with the goal role *Maintain*. The connectors *HeadOfChain* and *TailOfChain* denote the mutual situated commitments of two agents that collaborate

[1] To allow agents to set up a chain of more than two agents, an additional role *Link* would be necessary.

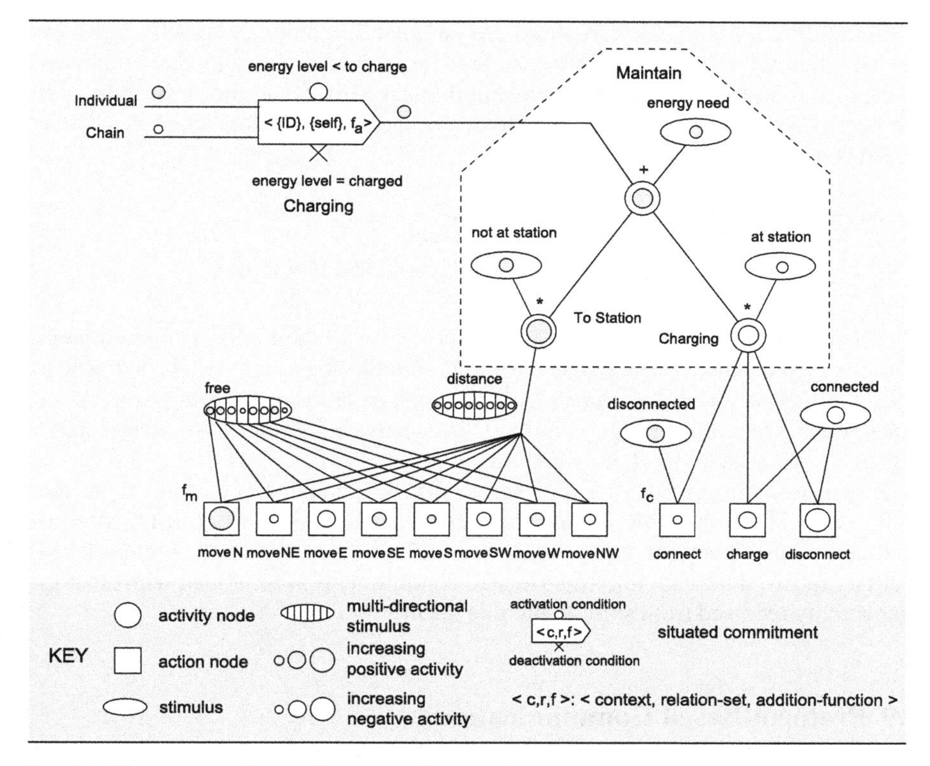

Fig. 3.8 Situated commitment *Charging* with its goal role *Maintain*

to pass loads in a chain. These situated commitments connect the single root node of *Individual* with the route nodes of *Head* and *Tail*, respectively.

Figure 3.8 shows the situated commitment *Charging* together with its goal role *Maintain* in detail. Besides a name, each situated commitment is characterized by an *activation condition*, a *deactivation condition*, and a three-tuple ⟨*context, relation-set, addition-function*⟩. Activation and deactivation conditions are boolean expressions based on the agent's internal state. The activation condition for *Charging* in Fig. 3.8 is *energy level* < *to charge*, i.e., as soon as the energy level of the agent crosses the threshold *to charge*, the activation condition becomes true and the situated commitment is activated.

The relation set contains the identities of the agents involved in the situated commitment. The context describes contextual properties of the situated commitment such as properties of elements in the environment (e.g., the distance to particular objects in the environment). Since *Charging* is a commitment of the agent relative to itself, the relation set is {*self*}. The context of *Charging* is {*ID*}, the identifier of the charge station. For example, for an agent that commits to be *HeadOfChain* in a collaboration (see Fig. 3.7), the relation set is the agent that is *TailOfChain*, and the context contains the type of loads that are passed between the collaborating agents.

Finally, the addition function determines, when the commitment is activated, how the activities of the source roles are combined into a resulting activity that is injected in the goal role. When the *Charging* commitment is activated it injects an additional amount of activity in the *Maintain* role, determined by the addition function f_a. A possible definition for f_a is as follows:

$$A_{Charging} = A^{+}_{Individual} + A^{+}_{Chain}$$
$$\text{with } A^{+}_{Node} = A_{Node} \text{ iff } A_{Node} > 0\text{, and 0 otherwise}$$

The *Maintain* role combines the additional activity of the *Charging* commitment with the regular activity accumulated from its stimuli. The deactivation condition of *Charging* is *energy level = charged*, i.e., as soon as the accumulated energy level reaches the *charged* level the commitment is deactivated. Then *Charging* no longer influences the activity level of its goal role.

In general, an agent can be involved in different situated commitments at the same time. The route node of one role may receive activity from different situated commitments and may pass activity to different other situated commitments. Activity received through different situated commitments is combined with the regular activity received from stimuli into one result.

3.9 Protocol-Based Communication

3.9.1 Primary Presentation

The primary presentation of the protocol-based communication pattern is shown in Fig. 3.9. Protocol-based communication comprises three data repositories: Inbox, Outbox, and Conversations and five components: Message Receiving, Message Sending, Message Decoding, Message Encoding, and Communicating.

3.9.2 Architectural Elements

The *Conversations* repository maintains a set of *conversations*. A conversation is an ongoing communicative interaction following a well-defined *communication protocol*. A communication protocol consists of a series of protocol steps. Each protocol step is characterized by a condition–effect pair. The condition determines whether the step is applicable. Conditions take into account the agent's current knowledge and data from ongoing communicative interactions. The effect is the actual result of executing the protocol step (see below). A conversation is initiated by the initial message of a communication protocol. At each stage in the conversation there is a limited set of possible messages that can be exchanged. Terminal states determine when the conversation comes to an end.

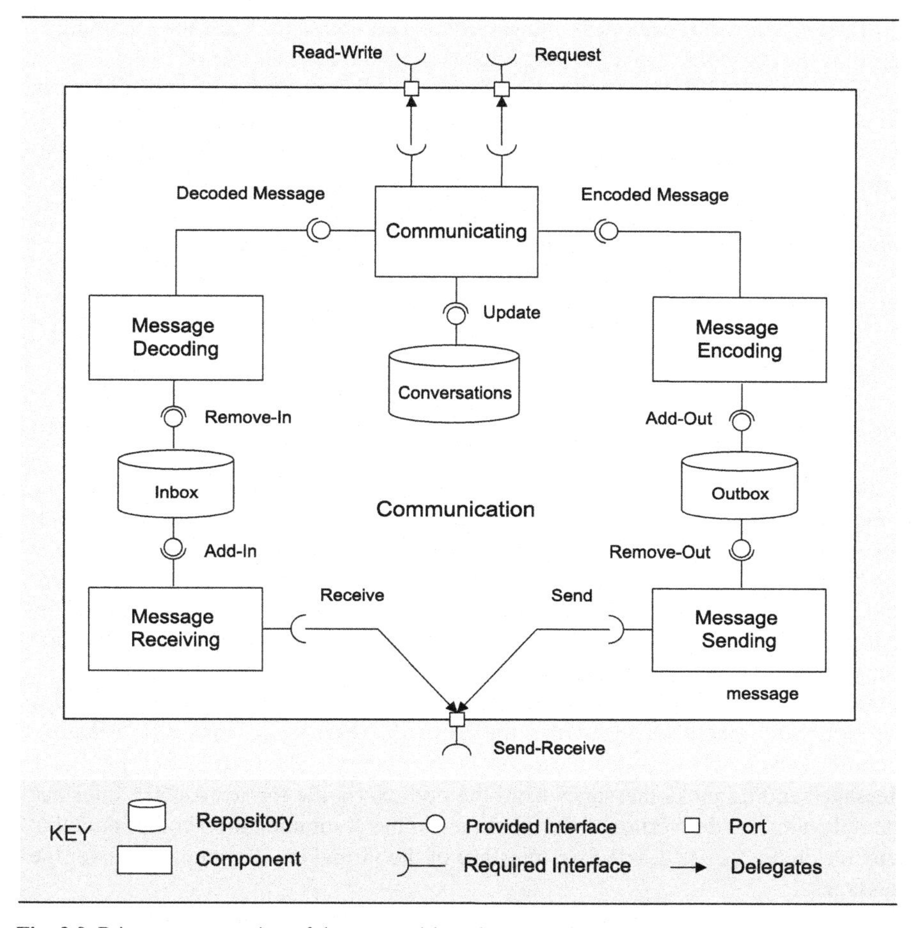

Fig. 3.9 Primary presentation of the protocol-based communication pattern

The *Communicating* component provides a dual functionality: (1) it interprets decoded messages and reacts appropriately; (2) it initiates and continues a conversation when the necessary conditions hold. During the execution of a protocol step, communicating may initiate a perception request. The execution of a protocol step will produce the data to encode a new message and update the corresponding conversation. Furthermore, the agent's current knowledge may be modified, possibly affecting the agent's selection of actions. A typical example is the activation and deactivation of a situated commitment.

Outbox and *Inbox* are message buffers. They buffer outgoing and incoming messages, respectively.

Message Sending selects a pending message from the outbox buffer and passes it to the communication service of the virtual environment. *Message Receiving* accepts messages from the communication service.

Message Encoding encodes a newly composed message. Message encoding is based on the *communication language* that is shared among the agents in the system. A communication language defines the format of the messages, i.e., the subsequent fields the message is composed of. The message content is based on an *ontology* that defines a shared vocabulary of words that agents use to represent domain concepts and relationships between the concepts. *Message Decoding* selects a received message from the inbox buffer and decodes the message according to the given communication language and ontology.

3.9.3 Interface Descriptions

The interface descriptions specify how the components of communication are used with one another.

The conversations repository exposes the `Update` interface that enables the communicating component to access and modify the agent's ongoing conversations. The communicating component delegates read or write requests for the agent's current knowledge to the `Read-Write` interface of the communication component. Perception requests are delegated to the required `Request` interface of the communication component that depends on the provided interface of the perception component with the same name.

To encode new messages the communication component depends on the message encoding component that provides the `EncodedMessage` interface. Message encoding puts newly encoded messages in the outbox via the `Add-Out` interface. Message sending picks messages from the outbox via the `Remove-Out` interface and delegates the delivering of the messages to the communication component that depends on the `Send-Receive` interface of the virtual environment to deliver the messages.

Communication delegates received messages to the message receiving component via the `Receive` interface. Message receiving puts the messages in the agent's inbox via the `Add-In` interface. The `Remove-In` interface provided by the inbox allows the message decoding component to decode incoming messages. The date of decoded messages is passed to the communication component via the `Decoded Message` interface.

3.9.4 Design Rationale

Direct communication allows situated agents to exchange information and set up collaborations. Coordination through message exchange is complementary to indirect coordination via marks in the virtual environment such as pheromone-based coordination. The various components in the communication component are assigned clear-cut responsibilities and coupling among components is kept low.

The selection of messages from the inbox buffer and outbox buffer can be defined according to the application requirements at hand. A simple policy is first-in-first-out. An advanced policy can take into account runtime information such as the content of the messages and the agents involved in the interaction.

Communication defined in terms of protocols puts the focus of communication on the relationship between messages. In each step of a communicative interaction, conditions determine the agent's behavior in the conversation. Conditions depend not only on the status of the ongoing conversations and the content of received messages but also on the actual conditions in the environment reflected in the agent's current knowledge, in particular the status of the agent's commitments. This contributes to the flexibility of the agent's behavior.

3.10 Summary

In this chapter, we have shown how proven domain expertise with multi-agent system engineering can be captured by means of a pattern language. The presented pattern language builds upon the foundation of two decades of research and expertise with engineering situated multi-agent systems and integrates our experiences with the design of practical situated multi-agent systems. The target domain of the pattern language is applications that are subject to highly dynamic and changing operating conditions making flexibility and openness primary requirements, and in which global control is hard to achieve demanding for decentralized control.

The pattern language ties five different patterns together. Situated agent and virtual environment are the central patterns of the pattern language. Selective perception, roles and situated commitments, and protocol-based communication zoom in on the three main concerns of a situated agent. For each pattern, we have provided a primary presentation that shows the constituent architectural elements of the pattern, a catalog that explains the responsibilities of the element, an interface description that specifies how the elements are used with one another, and a design rationale that explains the underlying design choices and the quality attributes associated with the pattern.

The pattern language provides an asset base architects can draw from during architecture-based design of situated multi-agent systems. In the next chapter, we show how we have used the pattern language during the design of a decentralized control architecture for an automated transportation system.

Chapter 4
Architectural Design of Multi-Agent Systems

Architectural design concerns the primary structures of a software system. Central in architecture design of a multi-agent system is the achievement of the system's quality attributes based on design decisions. To make design decisions, architects use established practices such as architectural patterns. To be effective, a software architecture must be properly documented. Architectural views provide a proven approach to document the structures of a complex software system. Documenting specific concerns of multi-agent systems such as roles, organizations, and interaction protocols may require dedicated notations.

In architecture-based design of multi-agent systems, we use a design method that is based on attribute-driven design (ADD) [173]. ADD is concerned with the high-level decomposition of a software system which is critical for satisfying the system's quality requirements. The output of ADD is the first levels of a module decomposition view and other views as appropriate. To document the views, we follow a method based on Views and Beyond [45]. The architecture description typically includes a module view that documents the system's principal units of implementation, a component-and-connector view that documents the system's units of execution, and the deployment view that documents the relationships between the system's software and its environment.

We start this chapter with a general introduction of designing of a multi-agent system architecture with ADD and documenting its views with Views and Beyond. Then, we explain the design of the case study and we present the main views of the architecture documentation. We only refer briefly to middleware support for distribution and a number of related coordination concerns including task assignment and collision avoidance. These concerns are discussed in detail in the following chapters. The chapter concludes with a summary.

4.1 Designing and Documenting Multi-Agent System Architectures

A multi-agent system is a system that is structured as a set of autonomous agents that are able to flexibly adapt their behavior to changing operating conditions. Individual agents have only limited knowledge and control over the system as a whole. To

D. Weyns, *Architecture-Based Design of Multi-Agent Systems*,
DOI 10.1007/978-3-642-01064-4_4,

achieve the overall system functionalities and qualities, agents interact and coordinate their behavior. Architectural design of a multi-agent system concerns the concrete specification of the top-level structures of the system in order to achieve the stakeholders' requirements. Structures include the primary structures of individual agents and the structures of organizations of agents.

ADD follows a recursive process that decomposes system elements by applying architectural approaches that satisfy its driving quality attribute requirements. A pattern language provides a powerful vehicle for constructing software architectures of multi-agent systems.

4.1.1 Designing and Documenting Architecture in the Development Life Cycle

Figure 4.1 shows the part of the software development life cycle where architecture design and documentation fit in.

Architectural design with ADD can start when the main architectural drivers are known which include functional and quality requirements. If a utility tree is available with a ranked set of quality attribute scenarios, the scenarios which are very important to the stakeholders and which have potentially a high impact on the architecture are candidate architecture drivers. In each step of ADD, an architectural element is refined based on an architectural approach that realizes a set of requirements.

The architecture documentation is mostly produced in an iterative fashion, intertwined with the design of the system. Initially, documentation typically consists of a set of cartoon-like diagrams. In later phases of the design process, the description

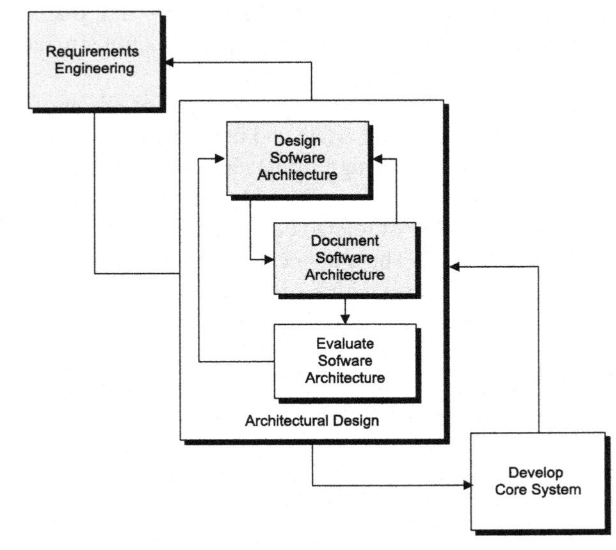

Fig. 4.1 Architecture design and documentation in the software development life cycle. *Shaded boxes* represent the activities of interest in this chapter

can be refined and rigorously documented. Views and Beyond offers templates to document the relevant views and additional documentation that applies to all views.

Architecture design ends when the software architect reaches a level of confidence that the foundation for the realization of the system requirements, in particular the quality requirements, is established.

4.1.2 Inputs and Outputs of ADD

The required inputs of architectural design with ADD are

- Functional requirements specify what functions a system must provide to meet the stakeholder requirements. An example of a functional requirement in the case study is that the battery of an AGV must be recharged when the energy is below a certain level.
- Design constraints are restrictions on the design that must be incorporated into the design of the system. For example, an 11 Mbps wireless LAN is available for communication.
- A set of prioritized quality attribute requirements that precisely specify the degrees to which a system must exhibit various quality properties. An example is the utility tree shown in Fig. 7.4.

The output of architectural design with ADD is a description of the software architecture of the system using various types of architectural views, including

- A set of module views that show the principle units of implementation, their responsibilities, and relationships.
- A set of component-and-connector views that shows the interactions among run-time elements and the properties of these interactions.
- A set of deployment views that show how the application software is allocated to computer hardware.

4.1.3 Overview of the ADD Activities

Figure 4.2 shows an overview of the activities of ADD [173].

At each stage of the decomposition, an architectural approach is chosen that satisfies a set of architectural drivers that are associated with the element that is decomposed. Next, functionality is allocated to the sub-elements. Finally, the decomposition is verified and the responsibilities with respect to requirements and constraints are assigned to the sub-elements. Using a pattern language to guide the design will affect the different selection criteria, including the selection of elements, drivers, and approaches to satisfy the drivers. The case study that follows illustrates the application of ADD with a pattern language for a concrete multi-agent system.

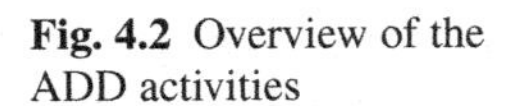

Fig. 4.2 Overview of the ADD activities

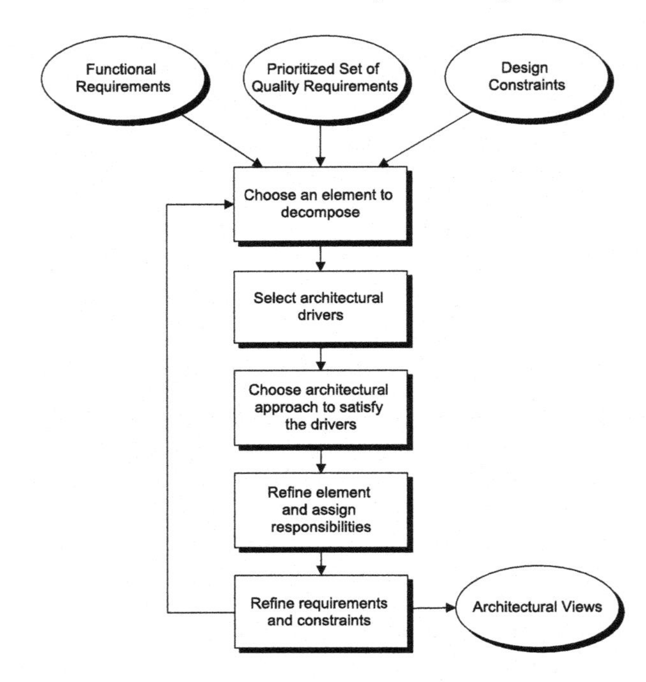

4.2 Case Study

The case study gives an extensive overview of the design and documentation of the software architecture for an AGV transportation system. We start this section with introducing the domain of automated transportation systems, we explain the business case, and we discuss the main functional requirements and quality requirements. In the following two sections, we give an overview of the architectural design of the agent-based system based on ADD and a detailed description of the multi-agent system architecture-based Views and Beyond.

4.2.1 The Domain of Automated Transportation Systems

Automatic guided vehicles (AGVs) are fully automated, custom-made vehicles that are able to transport loads in a logistic or production environment. An automated transportation system with AGVs can be used for distributing manufactured products to storage locations or as an inter-process system between various production machines. Figure 4.3 shows an AGV at work in a cheese factory.[1]

[1] http://www.egemin.com

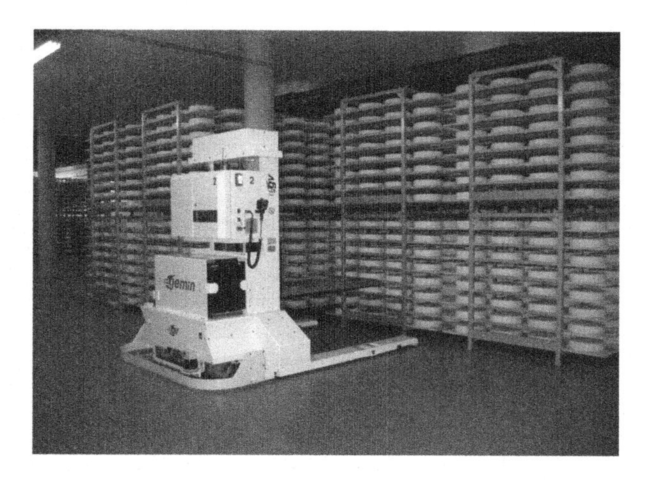

Fig. 4.3 An AGV at work in a cheese factory

Transports are generated by an enterprise resource planning (ERP) system and possibly operators and executed by AGVs. A transport includes picking up a load at a pick location, moving it to a drop location, and drop it there. An AGV is equipped with a steering system that provides the low-level control software connected to sensors and actuators to manipulate loads and move safely through the warehouse environment. While moving, the vehicles follow specific paths in the warehouse by means of a navigation system which uses stationary beacons in the work area such as laser reflectors on walls or magnet strips in the floor. Different navigation systems may be used in different sections of the warehouse (Fig. 4.4). While executing transports, AGVs may interact with machines. For example, an AGV may fetch a load from a rack or deliver a load onto a conveyor. To enable the AGVs to communicate with other systems, the warehouse provides a wireless LAN (local area network). In addition, stationary systems may use a wired LAN. AGVs are equipped with a battery as energy source that can be recharged at one of the available charging stations. When an AGV is idle, it can park at a free park location. Figure 4.5 summarizes the main concepts in the domain of AGV transportation systems.

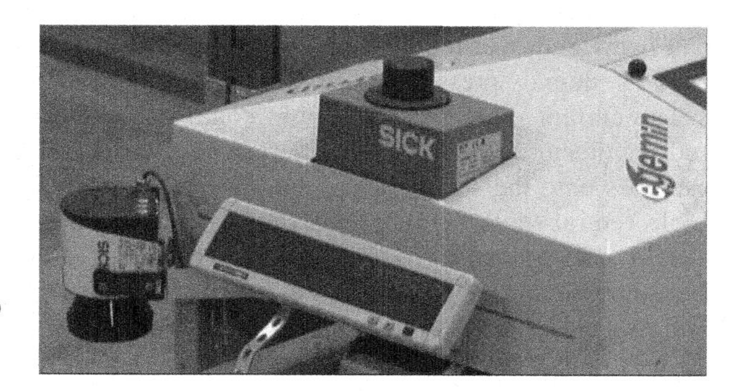

Fig. 4.4 Rotating laser scanner for navigation on top of the AGV. An obstacle detector in the front

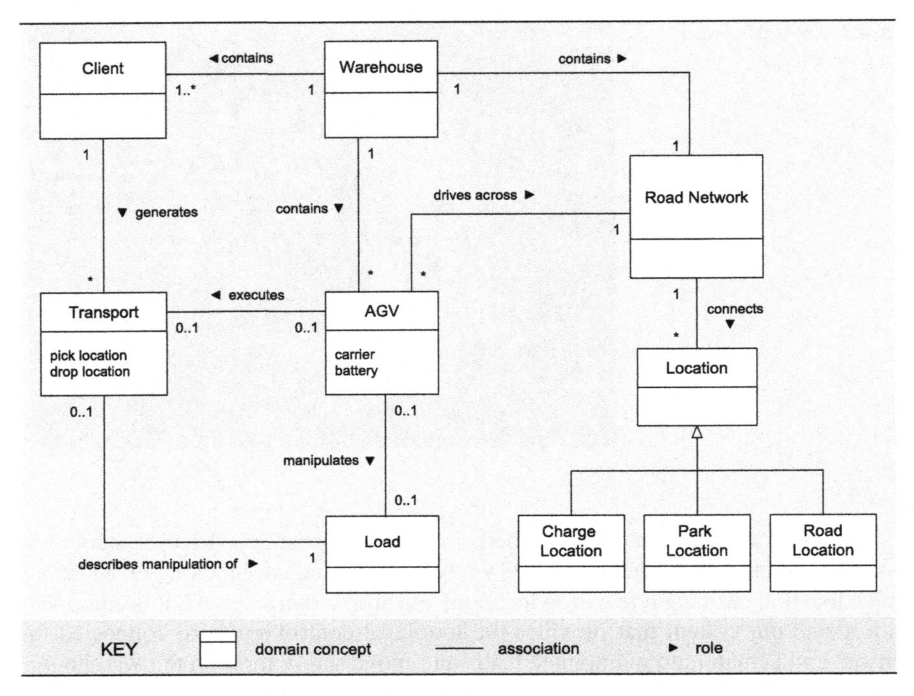

Fig. 4.5 Domain model with the main concepts of an AGV transportation system

4.2.2 Business Case

Egemin is an automation company that focuses on three areas: material handling, industrial automation, and consulting. In the material handling domain, the product flagship is the AGV transportation system. AGVs are controlled by traffic control software. The traffic control software is responsible for transport management and traffic control. Transport management is concerned with which loads have to be transported and where the loads have to be transported to. Traffic control is the process of routing AGVs efficiently while avoiding collisions and deadlocks between vehicles. Transport management should be organized in such a way that the workload is divided over the resources in an optimal way and that the capacity of the overall system (expressed in loads per hour) is at a maximum.

The current release of the traffic control software is conceived as a centralized architecture running on one server and scales up vertically by putting more resources onto the server (processing power, memory, etc.). The fact of having a single central server introduces a single point of failure. With the current release of the traffic control software, a customer project can be implemented efficiently, but the configuration of the rules the system has to comply to can be labor intensive, especially if the system has to be able to change its behavior according to changes in the environment. Thus, the flexibility of the system depends on the complexity of the rules configured and the experience of the engineer configuring those rules.

The objective of introducing a decentralized architecture is to scale up horizontally (by introducing more processors in a networked environment). Such an architecture aims at improved adaptability in response to the various dynamics in the warehouse environment and has no single point of failure. Instead of having one computer system that is in charge of numerous complex tasks, such as task assignment, routing, collision avoidance, and deadlock avoidance, in the new architecture the AGVs are provided with a considerable amount of autonomy. This opens perspectives to improve flexibility and openness of the system: the AGVs can adapt themselves to the current situation in their vicinity, task assignment is dynamic, and the system can deal autonomously with AGVs leaving and re-entering the system. The proposed architecture also strives to reduce the configuration effort. Hence a decentralized architecture based on agent technology becomes a logical next step. The resulting architecture should meet these business requirements and should also be efficient, robust, and easily deployable and maintainable.

4.2.3 System Requirements

We give an overview of the functionalities of the system and we discuss the main quality requirements. We also outline a number of important characteristics of industrial AGV transportation systems that have to be taken into account during architectural design. System requirements are kept fairly general, independent of any particular AGV system. In Chap. 7, we zoom in on specific functional requirements and quality attribute scenarios for a concrete AGV transportation application.

4.2.3.1 Main Functional Requirements

The main functionality the AGV transportation system has to perform is handling transports, i.e., moving loads from one place to another. A transport is composed of multiple jobs: a job is a basic task that can be assigned to an AGV. For example, picking up a load is a pick job, dropping it is a drop job, and moving over a specific distance is a move job. A transport typically starts with a pick job, followed by a series of move jobs, and ends with a drop job.

There should be enough AGVs available to execute the transports that enter the system, i.e., the AGVs should be able to handle the load of the system. In order to execute transports, the main functionalities the system has to perform are

1. Transport assignment: transports are generated by client systems and have to be assigned to AGVs that can execute them.
2. Routing: AGVs must route efficiently through the layout of the warehouse when executing their transports.
3. Gathering traffic information: although the layout of the system is static, the best route for the AGVs in general is dynamic and depends on the actual traffic

conditions and forecasts in the system. Taking into account traffic dynamics enables the system to route AGVs efficiently through the warehouse.
4. Collision avoidance: obviously, AGVs must not collide. AGVs cannot cross the same intersection at the same moment; however, safety measures are also necessary when AGVs pass each other on closely located paths.
5. Deadlock avoidance: since AGVs are relatively constrained in their movements (they cannot divert from their path), the system must ensure that AGVs do not find themselves in a deadlock situation.

To perform transport tasks, AGVs have to maintain their battery. AGVs can charge their battery at the available charging stations. Depending on the application characteristics, a vehicle recharges when its available energy is below a certain level or the vehicle follows a predefined battery charge plan or the vehicle can perform opportunity charging, i.e., the vehicle charges when it has no work to do. Finally, when an AGV is idle it can park at one of the available free park locations.

4.2.3.2 Main Quality Requirements

Stakeholders of an AGV transportation system have various quality requirements. *Performance* is a major quality requirement, customers expect that transports are handled efficiently by the transportation system. *Configurability* is important, it allows installations to be easily tailored to client-specific demands. Obviously, an automated system is expected to be *robust*, intervention of service operators is time consuming and costly.

Besides these "traditional" qualities, evolution of the market puts forward new quality requirements. Customers request for self-managing systems, i.e., systems that are able to adapt their behavior with changing circumstances autonomously. Self-management with respect to system dynamics translates to two specific quality goals: flexibility and openness.

Flexibility refers to the system's ability to deal with dynamic operating conditions autonomously. In the traditional centralized approach, the assignment of transports, the routing of AGVs, and the control of traffic are planned by the central server. The centralized planning algorithms applied by Egemin are based on predefined schedules. Schedules are rules associated with AGVs and particular locations in the layout, e.g., "if an AGV has dropped a load on location x, then that AGV has to move to node y to wait for a transport assignment." This approach lacks flexibility. A plan can only be changed under exceptional conditions. For example, when an AGV becomes defective on the way to a load, the transport can be re-assigned to another AGV. A flexible control system allows an AGV that is assigned a transport and moves toward the load to switch tasks along the way if a more interesting transport pops up. Flexibility also enables AGVs to anticipate possible difficulties. For example, when the amount of traffic is high in a certain area, AGVs should avoid that area; or when the energy level of an AGV decreases, the AGV should anticipate this and prefer a zone near to a charge station. Another desired property is that the system can handle particular situations autonomously, e.g., when a truck with loads arrives, the system should adapt its behavior taking into account this new task.

Openness of an AGV transportation system refers to the system's ability to deal autonomously with AGVs leaving and (re-)entering the system. Examples are an AGV that temporarily leaves the system for maintenance, and an AGV that re-enters the system after its battery is recharged. In some cases, customers expect to be able to intervene manually during execution of the system, e.g., to force an AGV to perform a particular job.

In summary, flexibility and openness are high-ranking quality requirements for today's AGV transportation systems. One possibility to tackle these new quality requirements would be to adapt the central planning approach aiming to improve the flexibility and openness of the system. By applying a situated multi-agent system, we investigated the feasibility of a radically new decentralized architecture to cope with the new quality requirements.

4.2.3.3 Specific System Characteristics

In addition to the functional requirements and quality requirements, a number of specific problem characteristics must be considered during architectural design:

- AGVs have to move toward loads before they can actually execute the transports. Moving toward a load may imply a considerable effort.
- AGVs are very constrained in their movements; they are confined to follow the paths of a predefined layout.
- The speed of AGVs is orders of magnitude lower than the speed of communication and the execution of the control software.
- A wireless LAN provides quasi-continual communication access to the distributed software system.

The architect has to take into account these problem characteristics when selecting suitable architectural approaches for the software architecture.

4.3 General Overview of the Design

In this section we give a general overview of the design of the multi-agent system for decentralized control of AGVs. We start with discussing a number of challenges we faced when applying multi-agent systems in practice. Next, we zoom in on the interaction of the system with its environment. Then, we briefly discuss the design process and we explain the rationale for the agent-based solution. Finally, we give a high-level overview of the architectural design of the multi-agent system.

4.3.1 Challenges at the Outset

From our experience, applying multi-agent systems in practice is a complex problem. During the design and development of the multi-agent system for AGV control, several difficulties were encountered. We explain the main challenges in turn.

Lack of Requirements Documentation. The general motivation to apply a situated multi-agent system for controlling AGVs was new quality requirements, in particular flexibility and openness. However, for a complex system such as the AGV transportation system, the stakeholders have various, often conflicting requirements. Unfortunately at the start of the project neither functional requirements nor quality requirements of the existing system were clearly documented. Information of the system was basically limited to user manuals. The absence of requirements documents resulted in contradictory opinions. This was further reinforced by the fact that some stakeholders had big, sometimes unrealistic expectations about agent technology, while others were more skeptic and showed some resistance to change.

Lack of Architectural Documentation. At the outset of the project, facing the complexity of AGV transportation system was overwhelming. A full-working AGV transportation system requires support for a variety of interdependent functionalities, including routing, collision avoidance, deadlock avoidance, interfacing with clients, task assignment, battery charging, and calibration of the vehicles. Complexity is further increased by the fact that different variants of functionalities are needed for different types of installations. Such complexity can only be managed through abstraction. Software architecture is centered on the idea of reducing complexity through abstraction and separation of concerns. Unfortunately the software architecture of the existing system was not documented. It turned out a difficult, time-consuming exercise to reconstruct the basic structures of the software architecture of the existing system. However, the reconstruction was crucial, not only to gain insight into how the system works but also to extract reusable parts of the existing code base.

Integrating a Multi-agent System with its Software Environment. In an industrial setting, systems are not built in isolation. When introducing a multi-agent system, it must be integrated with its environment (common frameworks, legacy systems, etc.). In Egemin, .NET is the standard environment and the company uses an in-house-developed component framework that provides common middleware services. Examples of legacy systems with which the multi-agent system needed to be integrated are the ERP system that generates the transport tasks and the steering system that provides the low-level control software of the AGVs. Dealing with these constraints raised severe challenges during the design and implementation of the multi-agent system.

Complementary Expertise. A particular challenge we faced with introducing multi-agent systems in practice came from the fact that two partners in the project had complementary expertise: an industrial partner with domain expertise and the academic partner with expertise in multi-agent systems. From the outset of the project, it was clear that the success of the project would depend on the mutual sharing of expertise and close, active cooperation between the two partners.

4.3.2 The System and Its Environment

We now zoom in on the interaction of the AGV transportation system with its environment. First, we discuss how the system interacts with external entities. Second, we explain how the agent-based control software is conceived as a layer on top of a common middleware platform for logistic systems that provides basic support for services such as persistency, security, and logging.

4.3.2.1 System Context Diagram

Figure 4.6 shows the context diagram of the AGV transportation system that describes how the system interacts with external entities. Transports are requested by client systems, i.e., an ERP system and possibly an operator. The AGV transportation system commands AGV machines to execute the transports, it monitors the status of the AGV machines, and it informs the clients about the progress of the transports. The transportation system can interact with external machines and command these machines to perform actions, e.g., opening a door or enabling a local conveyor element. Besides functionality to handle transports, the AGV transportation system provides a public interface to a monitor for observing the status of the logistic system. The monitor is an external software system that provides a graphical user interface allowing a user to follow the activity in the transportation system. Figure 4.7 shows a snapshot of the monitor. The monitor provides a real-time overview of the system. It visualizes a map of the warehouse layout with the moving AGVs. The monitor shows the pending, assigned, busy, and finished transports in the system, and it allows a user to inspect the status of transports and AGVs.

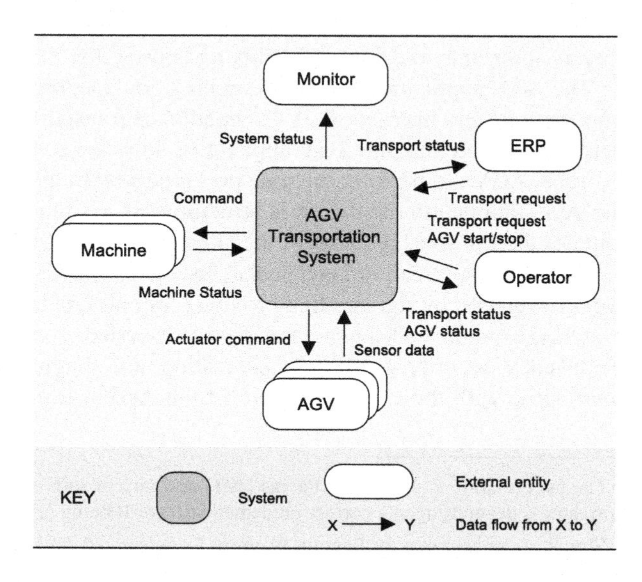

Fig. 4.6 Context diagram of the AGV transportation system

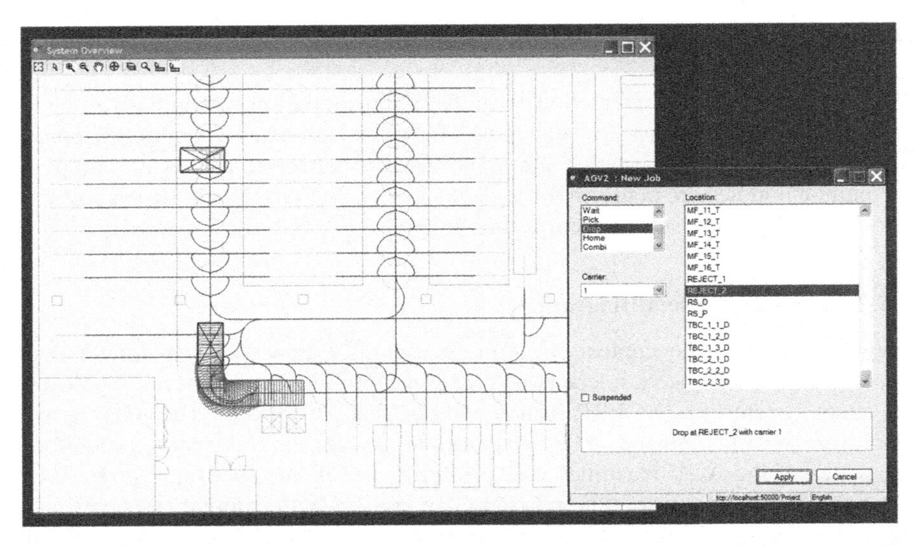

Fig. 4.7 Snapshot of the monitor. The *left hand side* shows a part of the warehouse layout with two AGVs. The *right hand side* shows a window that allows a user to add tasks manually

4.3.2.2 AGV Application and Supporting Middleware Services

Figure 4.8 shows a general overview of the software of the AGV transportation system. The software consists of three layers. Each layer provides a public interface with a cohesive set of services that other software can utilize without knowing how those services are implemented. The layers are allowed to interact with each other according to a strict ordering relation. In particular, a layer A is allowed to use[2] any of the public facilities of the virtual machine provided by the nearest lower layer B. Layers contribute to the modifiability and portability of a software system.

The AGV application layer is the application-specific software that accepts transport requests and instructs AGVs to handle the transports. In the traditional systems deployed by Egemin, the AGV application software consists of a central server that instructs AGVs to perform the transport requests. In the decentralized architecture, the AGV application software is structured as a situated multi-agent system that handles the transport requests of the clients.

The AGV application layer makes use of E'pia.[3] E'pia is a component framework developed by Egemin that provides common middleware services for logistic systems. E'pia provides general support for system configuration, communication, persistency, security, logging, visualization, and diagnosis. E'pia also handles the interfacing with the steering system of the AGVs. It translates high-level actuator

[2] The uses relation is defined by Parnas [121] as a unit of software A is said to use unit B if A's correctness depends upon a correct implementation of B being present.

[3] E'pia® is an acronym for Egemin Platform for Integrated Automation.

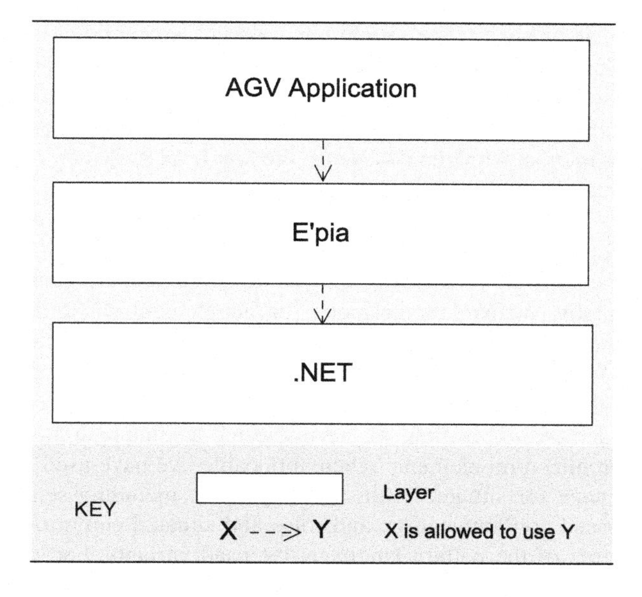

Fig. 4.8 Software layers of the AGV transportation system

commands to a low-level digital format of the actuator control software, and in the opposite direction, it parses the digital information derived from the sensors to provide a high-level representation of the actual status of the AGV.

The E'pia layer makes use of the Microsoft .NET framework [138]. The .NET framework provides a large body of pre-coded solutions to common program requirements, including support for user interfacing, database connectivity, network communication, and threading. .NET includes the Common Language Runtime environment (CLR) that serves as an application virtual machine shielding programmers from underlying platform details. The CLR also provides services such as security mechanisms, memory management, and exception handling.

The focus of the AGV transportation system described in this book is on the AGV application layer, i.e., on the decentralized control software composed of a situated multi-agent system.

4.3.3 Design Process

For the architectural design, we used ADD. Roughly spoken, the design process consisted of the following steps.

We started by using the basic patterns of the pattern language for situated multi-agent systems, situated agent and virtual environment, to map the basic system functionalities onto the basic components of the situated multi-agent system. The system comprises two types of situated agents, AGV agents and transport agents, that represent autonomous entities in the application. The virtual environment provides the means for agents to access resources, to exchange information, and to coordinate their behavior. The virtual environment is supported by the ObjectPlaces middleware that provides basic support for distribution and mobility. Simultaneously with the basic decomposition, we defined components for the interaction with external entities. In particular, we defined high-level components and interaction protocols to interact with client systems, the low-level AGV control software, and the monitor that allows remote inspection of the system.

Then, we have iteratively decomposed the agents and the virtual environment components. In each decomposition step, we selected an architectural element of the software architecture and we determined the target functional requirements and quality attribute requirements for that element. The order in which we have refined the architectural elements was essentially based on the incremental development of the application. We started with the functionality for one AGV to drive, then followed collision avoidance, next order assignment, deadlock avoidance, etc. For each decomposition, we have selected a suitable architectural pattern to refine the architectural element. When applicable, we have used a pattern of the pattern language for situated multi-agent systems, including selective perception, protocol-based communication, and roles and situated commitments. For some of the patterns of the pattern language, we used variants. For example, by using the same data representation in the virtual environment and the situated agents we could apply a number of simplifications for selective perception. For the design of the AGV agent, initially we used a free-flow tree with roles and situated commitments. With increasing complexity, we have redesigned the decision making component. We kept a free-flow tree for high-level decision making and added components to deal with specific aspects of decision making, such as task assignment and collision avoidance. For several specific functionalities, suitable architectural solutions had to be defined. For example, to support task assignment, routing, collision avoidance, and deadlock avoidance, we have developed appropriate coordination mechanisms. Therefore, we took inspiration from several well-known mechanisms for indirect coordination such as stigmergy and fields. In some cases, we have developed alternative solutions and performed a tradeoff analysis.

Architectural design ended when a suitable level of detail was reached to allow the developers to build the software. To validate the agent-based AGV transportation system, we used a setup with real AGVs, and tested in larger, industrially used simulations.

4.3.4 Design Rationale

The two main principles underlying the agent-based design of the AGV transportation system software architecture are control is decentralized and the architecture is structured as a situated multi-agent system.

First, control in the system is decentralized, i.e., no single central server controls (a large part of) the system. Decentralizing control implies local decision making. Local decision making supports self-management, i.e., entities can act locally and adapt their behavior to dynamics and disturbances in their local context. An advantage of decentralized control is increased reliability: there is no single point of failure. Furthermore, a decentralized architecture is more economical with respect to required processing power. Ideally, any central controlling processing unit is eliminated and all processing is moved to the nodes themselves. Since AGVs need to have processing units anyway, they are put to more use.

There are, however, some limitations and tradeoffs of decentralization [117]. First, the performance of the system may be affected by the communication links between nodes. Since more communication is typically needed, the communication infrastructure is more heavily loaded. Second, while the decentralized approach is designed to cope with disturbances, there is, in general, a tradeoff between its performance and the reactivity of the system to disturbances. A system that is designed to cope with many disturbances generally needs redundancy, usually to the detriment of performance and vice versa. Third, myopic decision may occur due to the lack of global information. While a central server has (more or less) a complete overview of the system, in a decentralized system such an overview does not exist. By using local information only, certain decisions for the system as a whole may be difficult to make or the decisions may lead to suboptimal solutions.

Architectural decisions are a tradeoff between trying to keep communication low, while being able to get the right information to the right place in a timely fashion. Since communication is the bottleneck, as a guideline communication it is kept local as much as possible.

As a second principle, the architecture is structured as a situated multi-agent system. The rationale behind this choice is the importance of the flexibility and openness requirements. Situated agents are autonomous entities that encapsulate their own state and behavior. A situated agent uses a behavior-based decision mechanism which guarantees responsiveness, robustness, and flexibility. Since each AGV, as an autonomous entity, acts locally, it can better exploit opportunities and adapt its behavior under changing circumstances. Situated agents work together to handle the stream of transportation tasks that enter the system. The agents can flexibly adapt their collaborations when the conditions in their vicinity change. Task assignment can be adapted when a new transport enters the system or when a more suitable AGV becomes available to perform the task. Situated agents commit to one another in a collaboration. However, the commitments are context dependable and as such alterable when circumstances ask for it.

4.3.5 High-Level Design

Now, we give an overview of the high-level design of the situated multi-agent system. We introduce the two types of agents that are used in the AGV transportation system, and we explain the structure of the virtual environment and show how agents use the virtual environment to coordinate their behavior. Then, we explain the basic programming abstractions of ObjectPlaces, views and roles, in some more detail and illustrate how they have supported the design of the situated multi-agent system.

Finally, we give a brief description of the AGV steering system that was fully reused in the decentralized control architecture.

4.3.5.1 AGV Agents and Transport Agents

We have introduced two types of agents: AGV agents and transport agents. The choice to let each AGV be controlled by an AGV agent is obvious. Transports have to be handled in negotiation with different AGVs, therefore we have introduced transport agents. An AGV agent is responsible to control its associated AGV. A transport agent represents a transport in the system and is responsible to ensure that the transport request is handled. Both types of agents share a common architectural structure that corresponds to the situated agent pattern of the pattern language for situated multi-agent systems. Yet, the two agent types have different internal structures that provide the agents with different capabilities.

AGV Agent. Each AGV in the system is controlled by an AGV agent that resides on a computer system located at the vehicle. The AGV agent is responsible for obtaining and handling transports and ensuring that the AGV gets maintenance on time. As such, an AGV becomes an autonomous entity that can take advantage of opportunities that occur in its vicinity and that can enter and leave the system without interrupting the rest of the system.

Transport Agent. Each transport in the system is represented by a transport agent. A transport agent is responsible for assigning the transport to an AGV and reporting the status and completion of the transport to the client that has requested the transport. Transport agents are autonomous entities that interact with AGV agents to find suitable AGVs to execute the transports. Transport agents reside at a *transport base*, i.e., a dedicated computer located in the warehouse.

Situated agents provide a means to cope with the quality attributes flexibility and openness. Particular motivations are (1) situated agents act locally; this enables agents to exploit opportunities and adjust their behavior with changing circumstances in the system and its environment—this is an important property for flexibility; (2) situated agents are autonomous entities that interact with one another in their vicinity; agents can enter and exit each other's area of interaction at any time—this is an important property for openness.

4.3.5.2 Local Virtual Environments

To realize the system requirements, AGV agents and transport agents have to coordinate. Agents have to coordinate for routing, for transport assignment, for collision avoidance, etc. One approach is to provide an infrastructure for communication that enables the agents to exchange messages to coordinate their behavior. Such approach, however, would put the full complexity of coordination in the agents resulting in complex architectures of the agents, in particular for the AGV agents. We have chosen for a solution based on the virtual environment pattern. The virtual environment pattern enables indirect coordination among the agents providing a separation of concerns that helps to manage complexity. Besides, the virtual

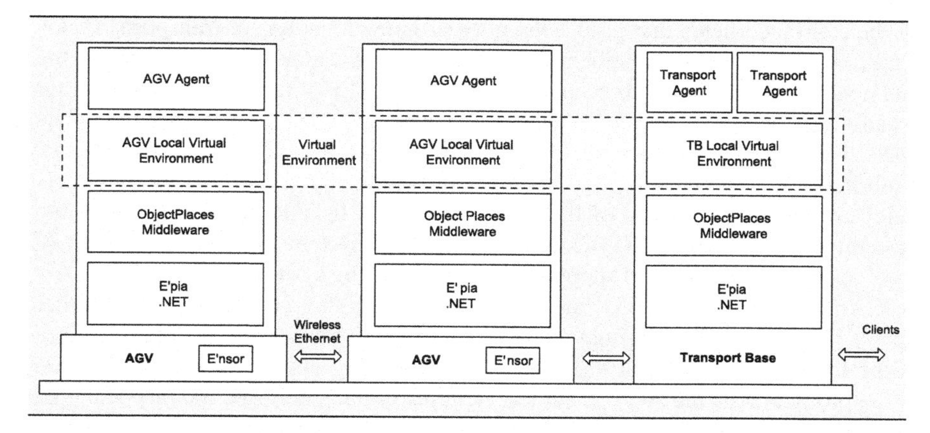

Fig. 4.9 High-level model of an AGV transportation system

environment serves as a suitable abstraction that shields the agents from low-level issues, such as the transmission of messages and the physical control of AGVs. Figure 4.9 shows a high-level model of an AGV transportation system.

Since AGV agents and transport agents are deployed on different nodes, the AGV and the transport base maintain a local virtual environment. The states of the local virtual environments are synchronized opportunistically, as the need arises. We explain state synchronization of local environments below when we elaborate on the ObjectPlaces middleware. The local virtual environments deployed on the nodes in the system are tailored to the type of agents deployed on the nodes. For example, the AGV local virtual environment deployed on the AGVs provides a high-level interface that enables the AGV agent to read out the status of the AGV and send commands to the vehicle. Obviously, this functionality is not available on the TB local virtual environment that is deployed on the transport base.

Coordination Through the Local Virtual Environment. The local virtual environment offers high-level primitives to agents to perform actions, perceive their neighborhood, and communicate with other agents. We illustrate with examples how the agents exploit the local virtual environments to assign tasks and to avoid collisions.

Transport Assignment. We have developed two approaches for adaptive transport assignment and used it in the AGV transportation system. FiTA (field-based transport assignment) is a field-based approach in which agents emit fields in the local virtual environment that guide idle AGV agents to loads. DynCNET is a protocol-based approach that extends standard contract net (CNET [151]). In DynCNET, the agents use explicit negotiation to assign transports. Here we illustrate FiTA in which agents coordinate through the local virtual environment. The basic idea of FiTA is to let each idle agent follow the gradient of a field that guides it toward a task that has to be executed. In FiTA, two types of fields are used: transport fields which are emitted by transports and AGV fields emitted by AGVs. Transport fields attract idle AGVs. However, to avoid multiple AGVs driving toward the same transport, AGVs emit repulsive fields. AGV agents combine perceived fields and follow the gradient of the combined fields that guides them toward pick locations of transports. Fields have a certain range and contain information about the source agent. The fields of the AGV agents have a fixed range and contain the identity of the AGV and its current location. The range of transport fields is variable and depends on the priority of the tasks. The spreading of the fields is a responsibility of the local virtual environments. With FiTA, the agents continuously reconsider the situation and task assignment is delayed until the execution of the task starts which benefits the flexibility of the system. When a task or AGV enters or leaves the system the perceived fields of local agents will be adapted supporting openness of the system.

Collision Avoidance. AGV agents avoid collisions by coordinating with other agents through the local virtual environment. AGV agents mark the path they are going to drive in their local virtual environment using *hulls*. The hull of an AGV is the physical area the AGV occupies. A series of hulls describe the physical area an AGV occupies along a certain path. If the area is not marked by other hulls (the AGV's own hulls do not intersect with others), the AGV can move along and actually drive over the reserved path. In case of a conflict, the involved local virtual environments use the priorities of the transported loads and the vehicles to determine which AGV can move on. AGV agents monitor the local virtual environment and only instruct the AGV to move on when they are allowed. Afterward, the AGV agents remove the markings in the local virtual environment.

These examples show that the local virtual environment serves as a flexible coordination medium: agents coordinate by putting marks in the local virtual environment and observing marks from other agents. We discuss collision avoidance and field-based task assignment in detail in Chaps. 5 and 6, respectively.

4.3.5.3 ObjectPlaces: Middleware for Mobile Applications

The mobility of the AGVs imposes highly dynamic operating conditions and inherent distribution of resources. A typical approach in mainstream software engineering is to support distribution with a suitable middleware. We have developed a middleware for mobile applications called ObjectPlaces. Mobile applications such as an AGV transportation system are characterized by (1) their need to take into account their physical environment (usually called context) explicitly and (2) their need to deal with dynamics and unexpected events originating from their context. ObjectPlaces proposes two new programming abstractions, views and coordination roles, to support mobile application development with respect to those two needs.

Views. The first abstraction, a view, is an up-to-date collection of data gathered from nodes in the network. The data is gathered from a number of data containers, called objectplaces, in which application components can share application-specific data objects to be viewed. An application component can specify which data to gather from objectplaces on which nodes use a constraint on properties of the nodes and a constraint on properties of the data objects. Based on these constraints, the middleware gathers the data objects in the view from remote objectplaces and keeps the data objects up-to-date with respect to changes in the properties of the nodes and changes in the remote objectplaces. The middleware encapsulates a view management service that builds and maintains views for the application.

The goal of views is to automate gathering application-specific information from application components on other nodes and to allow an application component to be aware of changes in this information. Using views, application components can coordinate by sharing information in objectplaces for others to view and by changing their behavior in response to information gathered from other application components.

In the AGV application, views are used for the synchronization of state of the local virtual environments on neighboring nodes. This synchronization is important for various coordination purposes. For example, a view is used to collect the candidate AGVs that are within a range of interest for a transport agent. Another example is the use of a view that maintains the AGVs in collision range of a particular AGV. The middleware builds and maintains the views according to the constraint defined for the view. For example, when an AGV approaches from a certain distance, it comes in collision range with an AGV. The middleware then includes that AGV in the view. Similarly, when an AGV leaves the collision range, that AGV will be removed from the view. As such, the application components have an up-to-date view of the AGVs in collision range that they can use to coordinate the vehicles avoiding collisions.

Coordination Roles. The second abstraction, a coordination role, encapsulates the behavior of an application component engaging in a protocol. A protocol is executed by two or more application components, each playing a particular coordination role. We call a particular exchange of messages by a particular group of application components using a protocol an interaction session. Such an interaction session needs to be started between a number of different roles, played by application components on distinct nodes. The middleware supports the setup and maintenance of such an interaction session, by managing the activation and deactivation of roles.

An application component that needs to start an interaction session, does so by indicating which coordination role it wants to play in the interaction session and by specifying with which other coordination roles on which other nodes it needs to interact. The nodes on which these coordination roles should be activated are specified declaratively by a constraint on properties of the nodes. The middleware activates the coordination roles on all nodes whose properties satisfy the constraints and that have an application component that can play the coordination role. The roles execute the protocol on behalf of the application components. The middleware encapsulates a coordination role activation service that monitors for changes in the network that cause a change in the composition of the group of interacting coordination roles, notifies the interacting roles if such a change occurs, and generally manages the activation and deactivation of coordination roles. The management of the interaction partners is thus handled by the middleware.

The goal of coordination roles is to support protocol-based coordination in mobile networks, by automating the setup of an interaction session between a group of roles, and maintaining the group of roles in the face of dynamics in the network.

For example, to avoid collisions at an intersection, AGVs need to execute a negotiation protocol with the group of all vehicles at the intersection, to determine which one can cross first. This group is dynamic, since AGVs arrive at and leave from the intersection continuously. The middleware automates the process of discovering the group of AGVs that are at the intersection, and maintaining the group of interacting vehicles as they arrive and leave.

Node Constraint. The declarative specification of a group of nodes to interact with, called the node constraint, is the key underlying idea of both abstractions. A node constraint specifies a group of nodes in the network by dictating a constraint on the node's properties relative to the node that starts the group. For example, an AGV can find a group containing all AGVs within 20 m or a group of AGVs within 30 m that are idle. Enabling the selection of interaction partners based on their properties has two main advantages. First, it allows more uncoupled interactions that can be tied easily to context properties such as location and status of the nodes. Second, the node constraint allows the encapsulation of the management of the interaction partners in the middleware. Since the interaction partners are frequently changing due to the mobility and overall dynamics in the network, this relieves the application developer of a tedious job.

4.3.5.4 Low-Level Control of AGVs with E'nsor

AGVs are equipped with a steering system that is called E'nsor.[4] We fully reused the steering system in the project. E'nsor provides an interface to command the AGV machine and to monitor its state. E'nsor is equipped with a map of the factory floor. This map divides the physical layout of the warehouse into logical elements: segments and nodes. Each segment and node is identified by a unique identifier. A segment typically corresponds to a physical part of a path of 3–5 m. E'nsor is able to steer the AGV per segment of the warehouse layout, and the AGV can stop on every node, e.g., to change direction. E'nsor understands basic actions such as `Move(segment)` that instructs E'nsor to drive the AGV over the given segment, `Pick(segment)` and `Drop(segment)` that instructs E'nsor to drive the AGV over the given segment and to pick/drop the load at the end of it, and `Charge(segment)` that instructs E'nsor to drive the AGV over a given segment to a battery charge station and start charging batteries there.[5] The physical execution of these actions, such as staying on track on a segment, turning, and the manipulation of loads, is handled by E'nsor. Reading out specific sensor data such as the current position and the battery level is also provided by E'nsor. The local virtual environment uses E'nsor to regularly poll the vehicle's current status and adjust its own state appropriately. For example, if the AGV's position has changed, the representation of the AGV position in the local virtual environment is updated.

---

[4] E'nsor® is an acronym for Egemin Navigation System On Robot.

[5] Actually, the instructions provided by the E'nsor interface are coded in a low-level digital format. The translation of actions to E'nsor instructions is handled by the local virtual environment.

4.4 Architecture Documentation

In this section, given an overview of the software architecture documentation of the AGV transportation system. The documentation follows the Views and Beyond approach. We start with a brief introduction in which we explain the different types of views of the documentation. Next, we zoom in on the different views of the software architecture.

4.4.1 Introduction to the Architecture Documentation

We start by introducing the types of views that comprise the software architecture documentation of the AGV transportation system. We briefly introduce the views and explain the mapping between the views. Then, we explain how each view is conceived as number of related view packages and we describe the template we use to document a view package.

4.4.1.1 Views and Mapping Between the Views

The software architecture of the AGV transportation system consists of three view types that highlight different system elements and different relationships and expose different quality requirements of the system. The allocation view type documents the relationships between the system's software and its execution environment. For this view type, the documentation provides the deployment view that shows how software elements are allocated to hardware. The module view type documents the system's principal units of implementation. For this view type, the documentation provides the uses view that defines depends-on relationships between modules. Finally, for the component-and-connector view type that documents the system's units of execution, the documentation provides the collaborating components view that shows how components collaborate to achieve required system functionalities.

Views describe a system or parts of it from a particular perspective. Although views focus on different aspects of the system, various elements and structures that appear in different views are related to one another. The mapping between views shows what pair wise view combinations have a mapping. Figure 4.10 summarizes the main mappings between the views of the AGV transportation system.

The annotation of the arrows give a general explanation of the relationships between the corresponding views.

4.4.1.2 View Package Template

Each view of the software architecture documentation is presented as a number of related view packages. A view package is a small relatively self-contained bundle of information of the system or a part of the system. The documentation of a view packet is organized in four parts as follows:

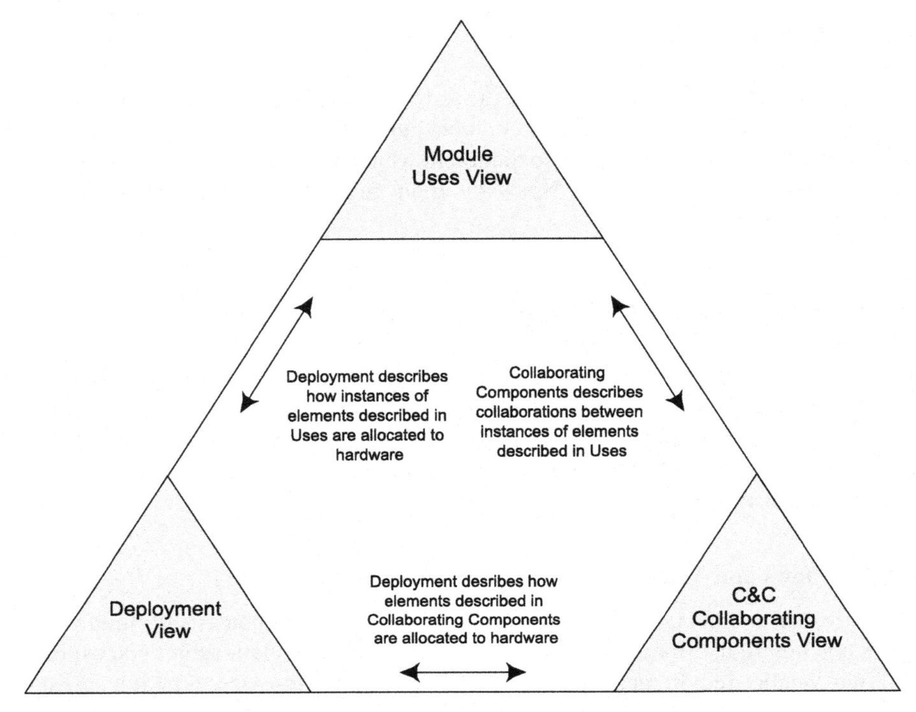

Fig. 4.10 Mapping between views

1. The primary presentation shows a graphical representation of the elements and their relationships in the view packet. A key explains the meaning of each symbol.
2. The element catalog describes the elements in the view packet and their properties. The element catalog also details the relations between elements and their properties.
3. Optionally, the element catalog specifies additional properties of elements and their relationships, such a detailed description of a particular element or a specification of the behavior of particular elements.
4. The architecture rationale explains the motivation for the design choices that were made. Rejected alternatives may be provided with a motivation why they were not chosen.

4.4.2 Deployment View

The deployment view is a style of the allocation view type that shows how the system software is allocated to hardware units. The elements of the deployment view are *software elements* and *environmental elements*. Software elements are usually components of the component-and-connector views. Environmental elements are computing hardware, including processors, data stores, network infrastructure. The relation between the elements is *allocated to* showing on which hardware units the software is deployed. Software elements are allocated to hardware units that execute code, store, or transmit data.

The deployment view can be used for performance analysis. Processor units have provided properties (CPU properties, memory, etc.) that need to be matched with the required properties of the allocated software elements. The communication among deployable units on different processing elements is an important focus of performance analysis. Important properties of communication channels are bandwidth and reliability. Bandwidth expresses the network capacity to transfer data among processing nodes, directly affecting the performance of a software system. Reliability is related to the system's behavior in the face of failing processing elements and communication channels. The deployment view shows dependencies among architectural elements and how a system is able to degrade gracefully in the face of a failure.

The software architecture of the AGV transportation system provides one view packet of the deployment view that describes how the system software is allocated to computer hardware.

4.4.2.1 Primary Presentation

Figure 4.11 shows the primary presentation of the deployment view. The application software consists of two types of subsystems with different responsibilities in the transportation system: transport base system and AGV control system.

4.4.2.2 Elements and Their Properties

The *transport base system* provides the software to manage transports in the AGV transportation system. The transport base system handles the communication with the ERP system. It receives transport requests and assigns the transports to suitable AGVs, and it reports the status and completion of the transports to the ERP system. The transport base system executes on a transport base, i.e., a stationary computer. The transport base system provides a public interface that allows an external monitor system to observe the status of the AGV transportation system.

The *AGV control system* provides the control software to command an AGV machine to handle transports and to perform maintenance activities. Each AGV control system is deployed on a computer that is installed on a mobile AGV. AGV control systems provide a public interface that allows a monitor to observe the status of the AGVs, and let an operator take over the control of the vehicle when necessary.

Communication Network. All the subsystems can communicate via a wireless network. The ERP system and machine software interacts with the AGV transportation system via the wired network. To debug and monitor the system, AGVs and the transport base can be accessed remotely via an external monitor system.

The properties of the environmental elements, in particular the characteristics of the communication channels, are important for the performance of the system.

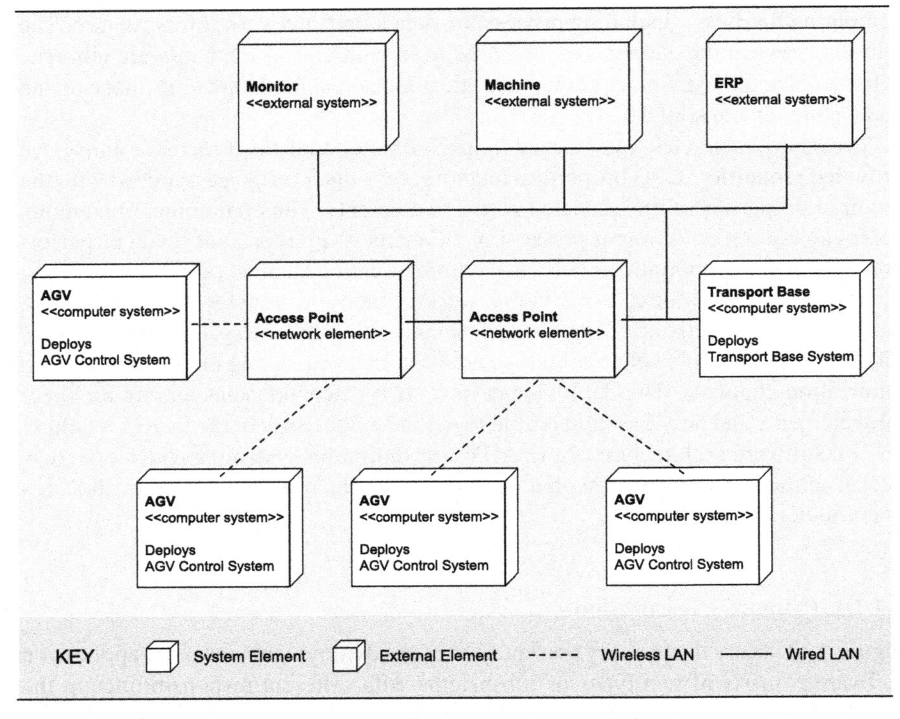

Fig. 4.11 Deployment view of the AGV transportation system

We discuss a performance analysis of a concrete AGV transportation system in Chap. 7.

4.4.2.3 Rationale

The top-level decomposition separates functionality for transport assignment (ensuring that the work is done) from functionality for executing transports (doing the work). The main motivations for the decomposition are the quality requirements flexibility and openness. By providing each AGV vehicle with an AGV control system, AGVs become autonomous entities that can take advantage of opportunities that occur in their vicinity and that can enter/leave the system without interrupting the rest of the system. Endowing AGVs with autonomy is a key property for adaptability of the system.

The separation of functionality for transport assignment and executing transports also supports incremental development. In the initial stage of the project, we developed a basic version of the AGV control system that provided support for performing transports and avoiding collisions. For test purposes, we manually assigned transports to AGVs. In the next phase, when we extended the functionalities of AGVs and integrated automated transport assignment, the top-level decomposition served as a means to assign the work to development teams.

4.4.3 Module Uses View

The module view type shows how a system is decomposed into manageable software units. The elements of the module view type are *modules*. A module is an implementation unit of software that provides a coherent unit of functionality. The basic criteria for module decomposition are the achievement of quality attributes. For example, changeable parts of a system are encapsulated in separate modules, supporting modifiability.

The module uses view is a style of the module view type. The relationship between modules in the module uses view is *uses*. A module uses another module if the correctness of the first module depends on the correct implementation of the second module. The module uses view documents how functionality is mapped to an implementation. It shows which implementation units use other units to achieve their functionalities. As such, the module uses view supports incremental development and is useful for debugging and testing of the system.

The software architecture of the AGV transportation system provides two view packets of the module uses view. We start with describing the primary decomposition of the AGV control system. Then, we show the decomposition of the transport base system.

4.4.3.1 AGV Control System

Figure 4.12 shows the primary presentation of the module uses view packet of the AGV control system.

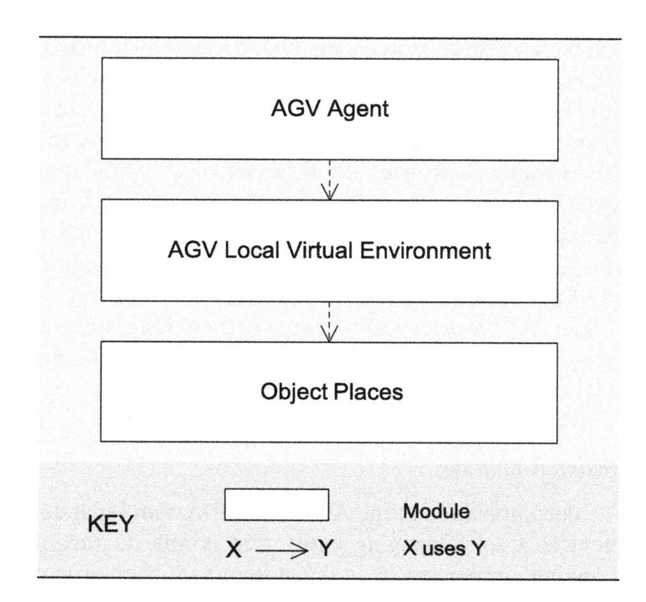

Fig. 4.12 Module uses view of the AGV control system

The basic structure of the AGV control system corresponds to the primary decomposition of a situated multi-agent system as explained in Chap. 3. The AGV agent is a situated agent that exploits the AGV local virtual environment to coordinate its behavior with other agents. The AGV local virtual environment is supported by the ObjectPlaces middleware that provides common services to deal with distribution and mobility.

Elements and Their Properties

AGV Agent. An AGV agent is responsible for controlling an AGV vehicle. The main responsibilities of the AGV agent are (1) obtaining transport tasks; (2) handling jobs and reporting the completion of jobs; (3) avoiding collisions; (4) avoiding deadlock; (5) maintaining the AGV machine (charging battery, calibrating, etc.); and (6) parking when the AGV is idle.

AGV Local Virtual Environment. The AGV local virtual environment offers a medium that the AGV agent can use to exchange information and coordinate its behavior with other agents. The AGV local virtual environment also shields the AGV agent from low-level issues, such as the communication of messages to remote agents and the physical control of the AGV.

Particular responsibilities of the AGV local virtual environment are (1) representing and maintaining relevant state of the physical environment and the AGV vehicle; (2) representing additional state for coordination purposes; (3) enabling the manipulation of state; (4) synchronization of state with neighboring local virtual environments; (5) providing support to signal state changes; (6) translating the actions of the AGV agent to actuator commands of the AGV vehicle; and (7) translating and dispatching messages from and to other agents.

ObjectPlaces Middleware. The ObjectPlaces middleware enables communication with software systems on other nodes. ObjectPlaces provides support for views and coordination roles in mobile networks. A view is a collection of data collected from neighboring nodes. Application components can specify which data to gather using a constraint on properties of the nodes and the collected data. A coordination role encapsulates the behavior of an application component in a protocol-based interaction. ObjectPlaces automates the setup and maintenance of interaction sessions between a group of application components participating in the protocol. Views and interaction sessions are maintained by ObjectPlaces in the face of dynamics of the network.

The AGV local environment uses the middleware services (1) to exchange messages with agents on other nodes and (2) to synchronize its state with the state of local virtual environments on neighboring nodes.

Design Rationale

The decomposition of the AGV control system separates responsibilities. The AGV agent is a self-managing entity that is able to flexibly adjust its behavior with changing circumstances in the system and its environment. The AGV local virtual environment provides an abstraction that allows the AGV agent to interact and coordinate its behavior with other agents in a way that is not possible in the physical environment. The ObjectPlaces middleware provides basis services for inter-node coordination, hiding the tedious management tasks of distribution in mobile systems. Separation of responsibilities helps to manage complexity. An alternative for indirect coordination through the local virtual environment is an approach where the functionality to control an AGV vehicle is assigned to an AGV agent only and where AGV agents coordinate through message exchange via a communication service. Such a design, however, would put the main part of the complexity of coordination in the AGV agent, resulting in a more complex architecture.

An instance of the local virtual environment module is deployed on each node in the AGV system. As such the local virtual environment has to maintain its state with the state of other local virtual environments. By defining appropriate views, ObjectPlaces maintains the sets of nodes of interest for the application logic. For example, to avoid collisions, a view is defined that keeps track of all the vehicles within collision range; we discuss collision avoidance in detail in Chap. 5. Whenever a vehicle enters or leaves this range, the ObjectPlaces middleware will notify the AGV local virtual environment about the change. By taking the burden of mobility, ObjectPlaces relieves the application developer from a tedious task.

4.4.3.2 Transport Base System

Figure 4.13 shows the primary presentation of the module uses view package of the transport base system.

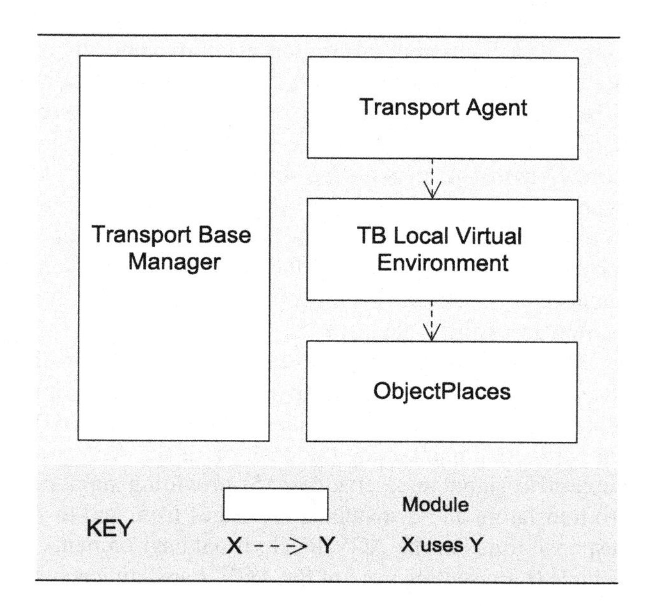

Fig. 4.13 Module uses view of the transport base system

The transport base system has a similar structure as the AGV control system and corresponds to the primary decomposition of a situated multi-agent system. The transport base manager accommodates the integration of the multi-agent system with clients.

Elements and Their Properties

The *transport base manager* has a dual responsibility. First, it is responsible for the communication with client systems, it accepts transport requests and reports the status of transports to clients. Second, it is responsible for creating transport agents, i.e., for each new transport request, the transport base manager creates a new transport agent. Each transport has a priority that depends on the kind of transport. The priority of a transport typically increases with the pending time since its creation.

A *transport agent* represents a transport in the system and is responsible for (1) assigning the transport to an AGV; (2) maintaining the state of the transport; and (3) reporting state changes of the transport to clients via the transport base manager. Physically, a transport agent is deployed on the transport base. Logically, however, the transport agent is located in the TB local virtual environment at the location of the load of the transport. For example, in Chap. 6, we will discuss a field-based approach for transport assignment in which a transport agent emits a field in the TB local virtual environment from the location of the load of the transport to attract idle AGVs.

The *TB local virtual environment* enables transport agents to coordinate with AGV agents to find suitable AGVs to execute the transports. Each transport agent has a limited view in the TB local virtual environment, i.e., each transport agent can only interact with the AGV agents within a particular range from its position. Yet, the range of interaction may dynamically change. In Chap. 6, we explain how the range of interaction of a transport agent dynamically extends when the agent does not find a suitable AGV to execute the transport. Limiting the scope of interaction is important to keep the processing of data under control.

Contrary to the AGV agents, the transport agents in the system share one local virtual environment. Still, the state of the TB local virtual environment has to be synchronized with the state of AGV local virtual environments in the system, e.g., to maintain the positions of the AGVs in the TB local virtual environment and the locations of new transports in the AGV local virtual environments. Since transport agents can access the TB local virtual environment concurrently, support for concurrent access is needed.

Particular responsibilities of the TB local virtual environment are (1) representing relevant state of the physical environment; (2) representing additional state for coordination purposes; (3) synchronization of state with AGV local virtual environments (in particular, maintaining the position of the AGVs in the system); (4) providing support to signal state changes; (5) providing support for concurrent access; and (6) translating and dispatching messages from and to AGV agents. Obviously, the responsibilities of the AGV local virtual environments that are related to the AGV vehicle (representing state of the AGV, translating actuator commands, etc.) are not applicable for the TB local virtual environment.

ObjectPlaces. The responsibilities of the ObjectPlaces middleware are similar as for the AGV control system, see Sect. 4.4.3.1.

Design Rationale

The transport base is in charge of handling the transports requested by the clients of the AGV transportation system. The transport base manager serves as an intermediary between the clients and the system. Apart from the transport base manager, the software architecture of the transport base is similar to the architecture of the AGV control system. Transport agents are situated in the TB local virtual environment that enables the agents to find suitable AGVs to perform the transport tasks. The ObjectPlaces middleware enables communication with the software systems on other nodes. The motivations for the decomposition of the transport base system are the same as for the AGV control system, see Sect. 4.4.3.1.

4.4.4 Collaborating Components View

A collaborating components view is a style of the component-and-connector view type. The collaborating components view shows a system as a set of interacting runtime components that use a set of shared data repositories to realize the required system functionalities. We have introduced the collaborating components view to explain how collaborating components realize various functionalities in the multi-agent system. The elements of the collaborating components view are components, data repositories, component–repository connectors, and component–component connectors:

- *Components* are runtime entities that achieve a part of the system functionality. Components are instances of modules described in the module view.
- *Data repositories* enable multiple components to share data.
- *Component–repository connectors* connect components with data repositories. These connectors determine which components are able to read and write data in the various data repositories of the system.
- *Component–component connectors* enable components to request each other to perform a particular functionality. Collaborating components require functionality from one another and provide functionality to one another.

The collaborating components view is an excellent vehicle to study the runtime behavior of a situated multi-agent system. The view shows the data flows between runtime components and the interaction with data stores, and it specifies the functionalities of the various components in terms of incoming and outgoing data flows.

The software architecture documentation provides two view packets of the collaborating components view. We start with the view packet that describes the collaborating components of AGV agent. Then follows the view packet that describes the collaborating components of AGV local virtual environment.

4.4.4.1 AGV Agent

This view packet zooms in on the software architecture of agents. We focus on the AGV agent. Figure 4.14 shows the primary presentation of the collaborating components view of the AGV agent. The structure of the AGV agent corresponds to the Situated Agent pattern, see Sect. 3.6.

Elements and Their Properties

The *current knowledge* repository contains state that the agent uses for decision making and communication. Current knowledge consists of static state and dynamic state. An example of static state is the value of LockAheadDistance. This parameter determines the length of the path AGVs have to reserve to drive smoothly and safely; we elaborate on path locking in Chap. 5. Examples of dynamic state are state collected from the observation of the AGV local virtual environment such as the

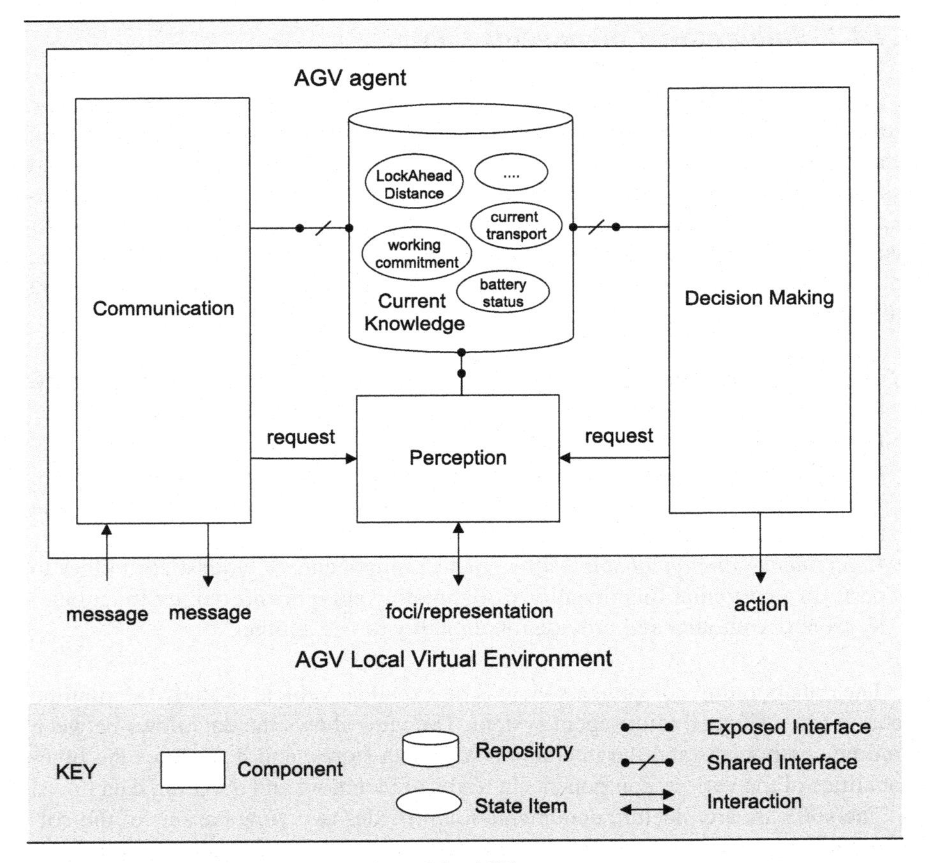

Fig. 4.14 Collaborating components view of the AGV agent

positions of neighboring AGVs, state related to ongoing collaborations with other agents, and runtime state related to the agent itself such as the battery status of the AGV. The current knowledge repository provides support for synchronized access. It offers a shared interface to the communication and decision making components that can concurrently read and write. The perception component is connected to a separate interface to update the agent's dynamic state according to the representations derived from observing the AGV local virtual environment.

Perception enables the AGV agent to sense the AGV local virtual environment according to the perception requests of communication and decision making and to update the agent's current knowledge accordingly. Requests are invoked by the decision making component and the communication component. Figure 4.15 shows the decomposition of the perception component. The structure of the perception component corresponds to the pattern of "Selective Perception," see Chap. 3. Sensing uses the set of selected foci to gather a representation of the AGV local virtual environment. The resulting percept is then filtered according to the set of selected filters of the perception request.

The *communication* component handles the communicative interactions of the AGV agent with other agents in the system. The main functionality of communication in the AGV transportation is handling messages to assign transports. Figure 4.16 shows the decomposition of the communication component. The communicating component is structured according to the "Protocol-Based Communication" pattern, see Chap. 3. Receiving accepts and buffers messages from the AGV local virtual environment. Decoding decodes the messages using the communication language, resulting in decoded message data. Communicating is the heart of the communication component. Communicating processes incoming decoded message data and produces outgoing encoded message data according to well-defined

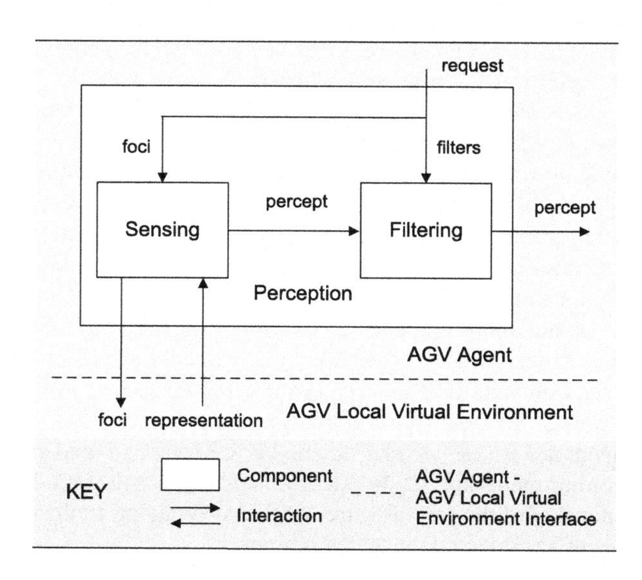

Fig. 4.15 Decomposition of the perception component of the AGV agent

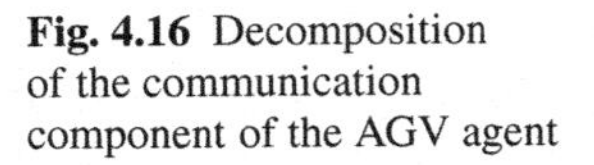
Fig. 4.16 Decomposition of the communication component of the AGV agent

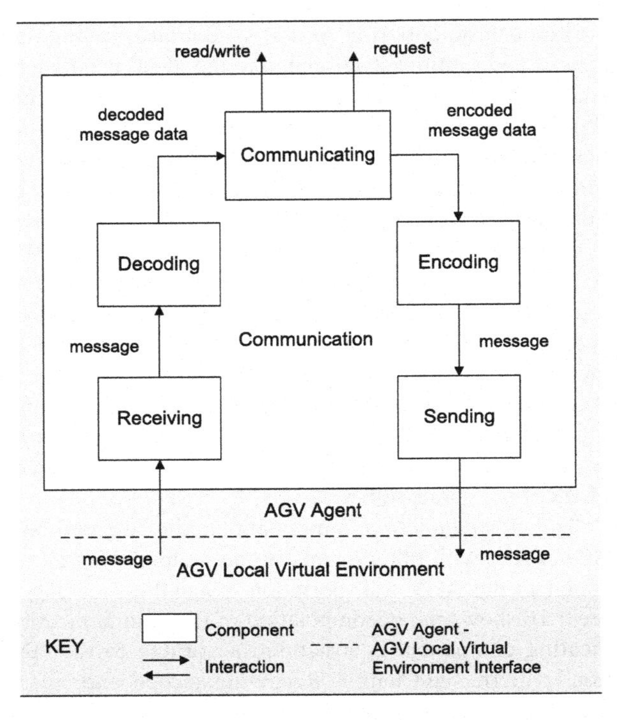

communication protocols. During the processing of a protocol step, the communicating component may invoke a perception request and read or write the agent's current knowledge. Encoding encodes newly composed message data. Sending collects the messages and passes the message to the communication service of the AGV local virtual environment. We zoom in on a concrete communication protocol for transport assignment in Chap. 6.

The *decision making* component handles the actions of the AGV agent. Due to the complexity of decision making of the AGV agent, we have modeled the decision making component as a hybrid architecture that combines a blackboard pattern with sequential processing. This architecture combines complex interpretation of data with decision making at subsequent levels of abstraction. The current knowledge repository serves as blackboard. Figure 4.17 shows the decomposition of the decision making component.

At the top level the action controller selects a high-level operator. Operator selection is designed as a free-flow tree extended with roles and situated commitments. The architecture of the tree has a structure similar to the one we discussed for the Roles & Situated Commitments pattern in Sect. 3.8.4. The main roles of the AGV agent are `role_working`, `role_charging`, and `role_parking`. The main commitments are `SC_WORKING` and `SC_CHARGING`. Figure 4.18 shows a runtime snapshot of the free-flow tree of the AGV agent. In the depicted situation, the AGV agent has selected the `PICK` operator.

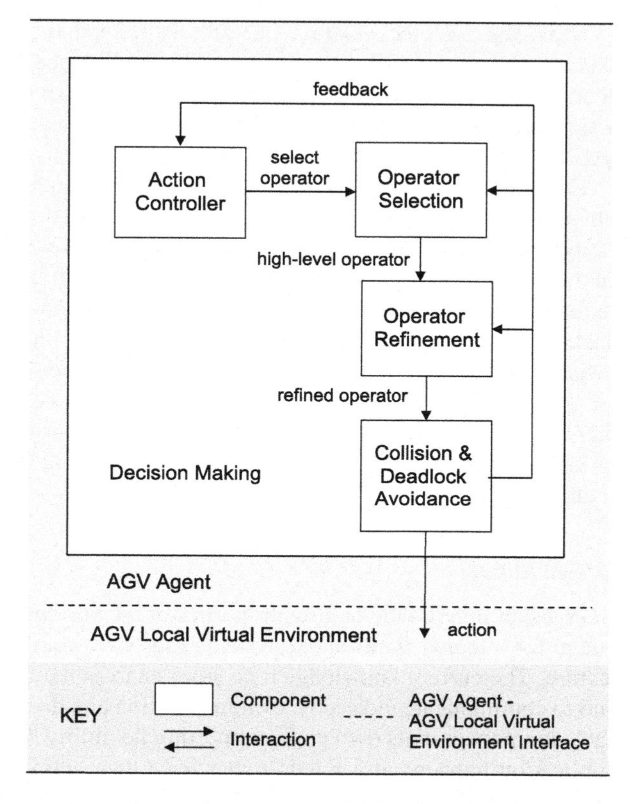

Fig. 4.17 Detailed structure of the decision making component of the AGV agent

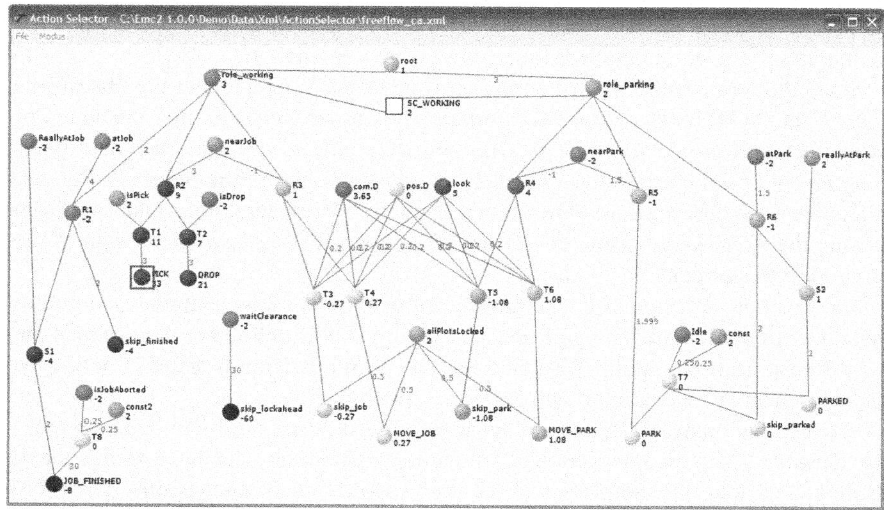

Fig. 4.18 Runtime snapshot of the free-flow tree of an AGV agent. The part of the tree that deals with battery charging is omitted

Next, the selected abstract operator is refined into a concrete operator. For example, when the pick operator was selected, the operator refinement component decides where the load has to be picked (`pick(segment x)`), or when a move action was selected, the component decides what segment is chosen to move on (`pick(segment y)`).

Finally, collision avoidance and deadlock avoidance are taken into account. Therefore, the trajectory associated with the selected operator is locked. As soon as the trajectory is locked, the selected action is invoked in the AGV local virtual environment. We discuss collision avoidance in depth in Chap. 5. If during the subsequent phases of decision making the selected operator cannot be executed, feedback is sent to the action controller that will inform the appropriate component to revise its decision. For example, if the operator refinement component has selected an operator `move(segment x)` and the collision avoidance module detects that there is a long waiting time for this segment, it informs the action controller that in turn may instruct the operator refinement component to consider an alternative route.

Design Rationale

AGV agent inherits the quality properties of the various patterns of the pattern language for situated multi-agent systems that were used to design the agent architecture. The current knowledge repository enables the data accessors to share state and to communicate indirectly. Communication and decision making act in parallel, each component in its own pace, supporting flexibility. Communication in the AGV application happens at a much higher pace than action selection. This difference in execution speed is exploited to continuously reconsider transport assignment in the period between an AGV starts moving toward a load and the moment when the AGV picks the load. A detailed discussion follows in Chap. 6.

Since the representation of the internal state of AGV agents and the observable state of the AGV local virtual environment are similar (for example, the status of the battery and the positions of AGVs), we were able to use the same data types to represent both types of state. As such, in comparison with the Selective Perception pattern, no descriptions were needed to interpret representations resulting from sensing the AGV local virtual environment. This resulted in a simple design of the perception component.

For an efficient design of the communication module, we have defined a domain-specific communication language and an ontology that is tailored to the needs of the AGV transportation system. We elaborate on communication in Chap. 6 when we discuss a protocol-based approach for task assignment.

In the initial phase of the project, we used a free-flow tree only for decision making. However, with the integration of collision avoidance and deadlock avoidance, it became clear that the complexity of the tree was no longer manageable. Therefore we decided to apply an architecture that allows incremental decision making. At the top level, a free-flow tree is still used to select an operator at a high level of abstraction; this preserves the advantage of adaptive action selection with a free-flow tree.

At the following levels, the selected operator is further refined taking into account collision avoidance and deadlock avoidance. Each component in the chain is able to send feedback to the action controller to revise the decision. This feedback loop further helps to improve flexible decision making.

4.4.4.2 AGV Local Virtual Environment

This view packet zooms in on the software architecture of the local virtual environment. We focus on the AGV local virtual environment. Figure 4.19 shows the collaborating components view of the AGV local virtual environment. The AGV local virtual environment is structured according to the "Virtual Environment" pattern, see Chap. 3.

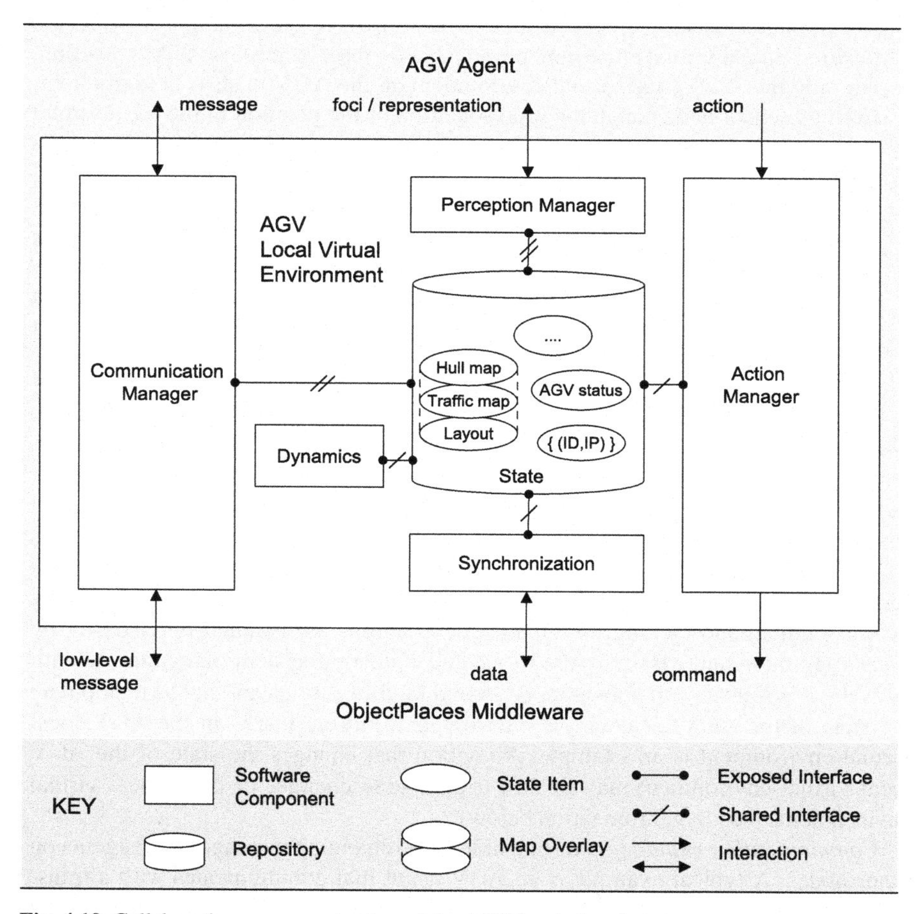

Fig. 4.19 Collaborating components view of the AGV local virtual environment

Elements and Their Properties

State. Since an instance of the AGV local virtual environment is deployed on each AGV in the system, each AGV local virtual environment is responsible for keeping its state synchronized with other local virtual environments. The state of the AGV local virtual environment is divided into three categories:

1. Static state: this is the state that does not change over time. Examples are the layout of the factory floor, which is needed for the AGV agent to navigate, and `(AGV id, IPnumber)` tuples used for communication. Static state must never be exchanged between local virtual environments since it is common knowledge and never changes.
2. Observable state: this is the state that can be changed in one local virtual environment, while other local virtual environments can only observe the state. An AGV obtains this kind of state from its sensors directly. An example is an AGV's position. Local virtual environments are able to observe another AGV's position, but only the AGV local virtual environment on the AGV itself is able to read it from its sensor and change the representation of the position in the local virtual environment. No conflict arises between two local virtual environments concerning the update of observable state.
3. Shared state: this is the state that can be modified in two local virtual environments concurrently. An example is a hull map with marks that indicate where AGVs intend to drive—we explain the use of hull maps in detail when we discuss collision avoidance in Chap. 5. When the local virtual environments on different machines synchronize, the local virtual environments must generate a consistent and up-to-date state in both local virtual environments.

Perception manager handles perception in the AGV local virtual environment. The perception manager's task is straightforward: when the agent requests a percept, for example, the current positions of neighboring AGVs, the perception manager queries the necessary information from the state repository of the AGV local virtual environment and returns the percept to the agent.

Action manager handles the actions of the AGV agent. The AGV agent can perform two kinds of actions. One kind is commands to the AGV, for example, moving over a segment and picking up a load. These actions are handled fairly easily by translating them and passing them to the AGV steering system that connects with the vehicle's sensors and actuators. A second kind of actions attempt to manipulate the state of the AGV local virtual environment. Putting marks in the AGV local virtual environment is an example. An action that changes the state of the AGV local virtual environment may in turn trigger state changes of other local virtual environments (see Synchronization below).

Communication manager is responsible for exchanging messages with agents on other nodes. A typical example is an AGV agent that communicates with a transport agent to assign a transport. Another example is an AGV agent that requests the AGV agent of a waiting AGV to move out of the way. The communication manager translates the high-level messages to low-level communication instructions that can be sent through the network and vise versa (resolving agent names to IP numbers, etc.).

Dynamics is responsible for maintaining dynamism in the AGV local virtual environment that happens independently from actions of agents or dynamics in the underlying environment. We give an example of such dynamism when we discuss the spreading of fields for field-based transport assignment in Chap. 6.

Synchronization has a dual responsibility. It periodically polls the status of the vehicle and updates the state of the AGV local virtual environment accordingly. An example is the maintenance of the actual position of the AGV. Furthermore, synchronization is responsible for synchronizing state with local virtual environments on other nodes. We explain the update process for collision avoidance in detail in Chap. 5.

Design Rationale

The AGV local environment component inherits the quality properties of the Local Virtual Environment pattern of the pattern language for situated multi-agent systems. Different functionalities provided by the local virtual environment are assigned to different components. This helps architects and developers to focus on specific aspects of the functionality of the local virtual environment. It also helps to accommodate change and to update one component without affecting the others.

A well-considered assignment of responsibilities among the main building blocks of the AGV control system, AGV agent, AGV local virtual environment, and ObjectPlaces, is crucial for managing complexity. For example, to avoid collisions, the AGV agent projects a hull in the AGV local virtual environment indicating its intended movement. The AGV local virtual environments of neighboring nodes resolve conflicts in case of a possible collision using a mutual exclusion protocol. The management of interaction partners that enter/leave the area and thus have to be included/excluded from the protocol is handled by the ObjectPlaces middleware.

Since an AGV agent continuously needs up-to-date data about the system (locations of the loads, status of hulls, etc.), we decided to keep the relevant state in the AGV local virtual environment synchronized with the actual state of the system. The synchronization component periodically polls the ObjectPlaces middleware to update the status of the system. As such, the state repository maintains an accurate representation of the state of the system to the AGV agent. As a result, the perception manager interacts only with the state repository, resulting in a simple design for perception management. In contrast, the representation generator in Virtual Environment pattern can also collect runtime data from the environment and integrate this data with local state of the AGV local virtual environment to produce a representation.

Changes in the system (e.g., AGVs that enter/leave the system) are reflected in the state of the local virtual environments, releasing agents from the burden of such dynamics. As such, the AGV local virtual environment, supported by the ObjectPlaces middleware, supports openness.

4.5 Summary

In this chapter, we discussed the design and documentation of multi-agent system architectures and we showed their use for a concrete multi-agent system.

In architecture-based design of multi-agent systems, we use ADD as a systematic method to design an agent-based system. ADD involves a recursive process in which system elements are decomposed by applying architectural approaches that satisfy its driving quality attribute requirements. The output of ADD is a set of views that describe the primary structures of a multi-agent system architecture. To document the views of an architecture, we use the Views and Beyond method. In Views and Beyond documenting a software architecture consists of two complementary activities: (1) documenting the relevant views and (2) documenting the information that applies to multiple views.

In the case study, we extensively discussed the architectural design and documentation of an AGV transportation system. We discussed the design process based on ADD and explained how we have used the pattern language for situated multi-agent systems to shape the software architecture of the AGV application. We motivated the main architectural decisions, including the use of AGV agents and transport agents, the use of local virtual environments that provide as a flexible coordination medium for the agents, and the supporting ObjectPlaces middleware that encapsulates the tedious management of distribution and mobility.

After the general overview of the design, we presented the various views of the software architecture documentation. The deployment view documents the allocation of the two subsystems, AGV control system and transport base system, to hardware. The module uses view specifies the responsibilities of the main modules of the subsystems and shows their dependencies. Finally, the collaborating components view provided insight into the internal structure and behavior of the AGV agent and AGV control system.

Chapter 5
Middleware for Distributed Multi-Agent Systems

One of the major challenges in the software development of a distributed multi-agent system is the coordination necessary to align the behavior of the agents. Since coordination determines whether agents cooperate effectively, it has a direct impact on the satisfaction of a distributed application's functional requirements. Furthermore, since coordination is realized primarily by communication, coordination has a large impact on quality attributes such as efficiency and resource usage.

Decentralization of control implies a style of coordination in which the agents cooperate as peers with respect to each other, and no agent has global control over the system or global knowledge about the system. As a result, complex interactions are necessary to achieve consensus since there is no single agent that can make a centralized decision. In the case of mobile applications, agents have to take into account the distribution of the nodes in physical space and other properties of the environment, which add extra complexity to the realization of coordination. Since development of distributed multi-agent systems is difficult, usually middleware is used to support the application developer.

We start this chapter with introducing middleware support for distributed systems and multi-agent systems in particular. Then, we explain in detail a concrete middleware that was developed for the case study and we illustrate how this middleware supported a complex coordination problem in a mobile setting. The chapter concludes with a summary.

5.1 Middleware Support for Distributed, Decentralized Coordination

We give an overview of the role of middleware for supporting the development of distributed systems. First, we zoom in on the multiple layers of middleware in distributed software systems in general. Then, we take a closer look on middleware for multi-agent systems.

D. Weyns, *Architecture-Based Design of Multi-Agent Systems*,
DOI 10.1007/978-3-642-01064-4_5, © Springer-Verlag Berlin Heidelberg 2010

5.1.1 Middleware in Distributed Software Systems

Over the last decade, the development of software systems increasingly emphasizes the reuse of software components. There is an ongoing trend away from programming applications from scratch to integrating them by configuring and customizing reusable components and frameworks [145]. Requirements for greater reuse in developing distributed software systems motivate the use of middleware-based architectures. Middleware is software that resides between the application and the underlying operating systems, network, and hardware. Middleware shields software developers from low-level tedious and error-prone platform details. It provides software developers with a consistent set of higher level abstractions and services closer to the application requirements. Figure 5.1 shows the multiple layers of middleware in distributed software systems [145].

Host Infrastructure Middleware encapsulates communication with the operating system. Widely used examples are the Java Virtual Machine and the .NET platform. Distributed Middleware defines higher level distributed programming models with reusable APIs and components that help programming distributed applications. Examples are Java Remote Method Invocation, Common Object Request Broker Architecture (CORBA), and SOAP that provide a simple XML-based protocol allowing applications to exchange structured information on the Web. Common Middleware Services define higher level domain-independent services that support

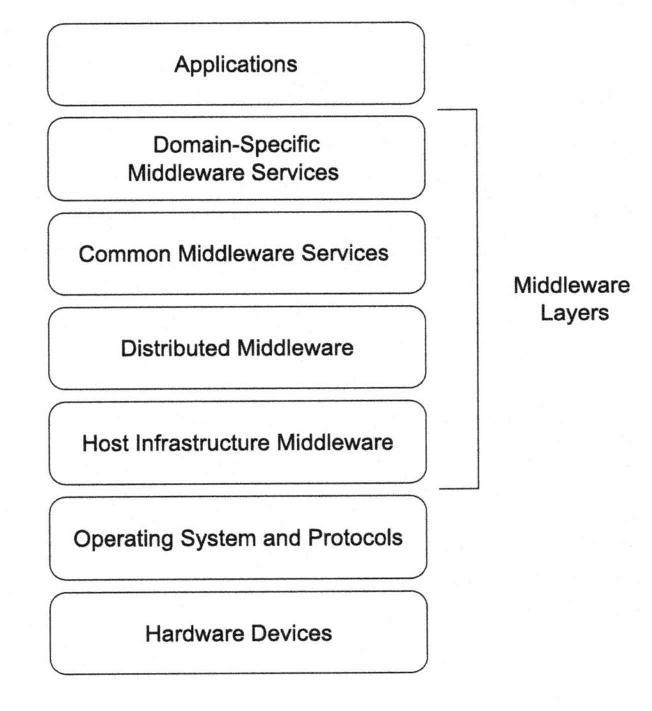

Fig. 5.1 Middleware layers with surrounding context [145]

programming of application logic such as transactional behavior, security, and database access. An example technology is Enterprise Java Beans that enables software developers to link predefined services ("beans") without having to write much code from scratch. Finally, Domain-Specific Middleware Services are tailored to the requirements of a particular interest group. Examples are middleware services for telecom, electronic commerce, and grid computing. Today, domain-specific middleware services tend to be less mature partly due to the lack of common middleware standards which are needed to provide a stable basis to create domain-specific services [145].

As distributed software applications have to deal with increasing dynamics and heterogeneity, software must be dynamically composed and adapted at runtime. A major trend in middleware is to combine domain-specific middleware functionality with specific component frameworks (e.g., JEE, .NET, etc.). This approach enables the construction of applications from independently developed third-party components and integrate built-in services covering nonfunctional requirements of a distributed application such as persistency and security. A typical example is service-oriented architectures [9] where the major part of application development boils down to assembling domain-specific services that comply with a set of declaratively specified policies. The complexity of flexible composition and runtime adaptation of services in the face of the crosscutting nature of functionality is the subject of active research [26, 95].

5.1.2 Middleware in Multi-Agent Systems

We now look at how typical middleware support for multi-agent systems maps on the different middleware layers:

- Distributed and host infrastructure middleware. Multi-agent system engineers generally consider distributed middleware services (RMI, CORBA, SOAP, etc.) as a basic platform to build multi-agent systems. The services provided by the bottom layer are not the main focus of research on agent environments but are typically considered as given infrastructure.
- Common middleware services. In multi-agent system development, common middleware services such as security, persistency, transactions are often considered minimally. For lab prototypes, there is a tendency not to consider these domain-independent services. Since the number of deployed multi-agent systems is rather limited, there is little experience with integrating common middleware services in multi-agent systems. Some platforms provide basic support for particular common middleware services such as Retsina [156] (security, monitoring, and logging) and Living Systems of Whitestein Technologies [171] (among others, transactions, persistency, and Web service access).
- Domain-specific middleware services. Support for agent interaction such as communication services for message exchange and infrastructures for coordination are part of the domain-specific middleware services layer. These infrastructures

are built on top of the distributed middleware platform and comprise programming abstractions and services that can be reused across multi-agent system applications [62]. Almost all agent platforms offer some form of domain-specific middleware service. The types of support are very different and include support for distributed message communication such as Jade [23], electronic institutions [54], artifacts [137], pheromone infrastructure [35], and infrastructures based on tuplespaces [113, 106]. Some examples of more specific approaches are delegate multi-agent systems [72], tag-based interaction [128], and communication filters [144].

Domain-specific middleware can help multi-agent application developers by simplifying and accelerating common development tasks [146]. Middleware simplifies application development by offering programming abstractions that hide lower level details from the application developer. It accelerates application development by encapsulating generic, reusable functionalities to support the programming abstractions. In particular, middleware encapsulates the tedious management tasks associated with distribution. As such, middleware offers conceptual and technical tools to support the application developer in dealing with the distributed aspect of the multi-agent system.

5.2 Case Study

The case study gives an extensive description of a domain-specific middleware for multi-agent systems and its application to the AGV transportation system. This middleware, called ObjectPlaces, supports the development of distributed, decentralized applications that are deployed in a mobile network. We start this section by characterizing the target systems of the middleware and derive requirements for the coordination middleware. Then, we introduce the basic building blocks of the middleware: objectplaces, views, and coordination roles. In the two following sections, we give a description of the software architecture of the middleware and explain how we have applied ObjectPlaces to solve the coordination problem of collision avoidance in the AGV transportation system.

5.2.1 Scope of the Middleware and Requirements

The ObjectPlaces middleware targets mobile applications with the following three characteristics:

1. *Context Awareness.* The applications have a strong connection with their context and actively need to take their context into account when coordinating. Typically, coordination solutions are expressed in terms of the current context properties of

application components,[1] in particular with respect to a components' interaction partners. For example, to execute a transport from a particular location, an AGV is selected among the AGVs within *a range of 30 m*.
2. *Dynamics.* The applications are subject to unexpected dynamics originating from the environment. These dynamics may be the result of the mobility of the nodes or of other changes in the application's context. As a result of dynamics, and the need for application components to be aware of changes in their context, application components need to be aware of the changes in interaction partners. For example, AGVs may move in and out of collision range of a particular intersection.
3. *Decentralization.* The applications we consider consist of distinct application components that cooperate as peers to reach the overall goal of the application. No single component has global control over or knowledge about the system. Decentralization of control typically increases both the importance and the complexity of coordination in the application.

These characteristics and the associated problems motivate the following requirements for middleware for mobile applications:

1. *Discovery of Interaction Partners by Properties.* Interaction partners should be discovered based on their properties, such as location of a node, status of the node. The identification of interaction partners should be expressed by using a declarative constraint on node properties.
2. *Management of Changes in Interaction Partners.* The supporting abstractions should allow the middleware to encapsulate the management of the group of components with which a particular component interacts, thereby removing this burden from the application developer.
3. *Decentralized Architecture of the Middleware.* The middleware should not introduce a centralized element in its architecture, as this would make the middleware unusable for decentralized applications.

In addition, the middleware should be efficient, i.e., it should consume a reasonable amount of bandwidth. We do not consider the middleware's overhead in computing space and time: bandwidth is the scarcest resource.

5.2.2 Objectplaces

Objectplaces are essentially containers of data objects. Objectplaces are not meant to be used by themselves, but the two main abstractions, views and coordination roles (explained in the following sections), are both used in conjunction with objectplaces. Hence, it is important to gain a basic understanding of objectplaces before explaining views and coordination roles.

---

[1] We use the term application component in its general meaning, i.e., a modular and independently describable entity that is part of an application. An AGV local virtual environment is an example of an application component.

5.2.2.1 Conceptual Model

An objectplace is a collection of objects that can be safely manipulated by concurrent processes using operations such as put and read and is as such a variant of a tuplespace [42]. The main motivation for developing a specific tuplespace variant is the need for asynchronous operations. Typical tuplespace operations are synchronous, i.e., a read operation reads a tuple from a tuplespace and blocks until the tuple is available. Due to the dynamic conditions in a mobile network, an asynchronous interface is needed. Objectplace operations return control to the caller immediately, and results are returned when they are available via a callback. This allows an event-driven style of interaction with the objectplace, which in the case of synchronous operations should be handled using polling.

Objectplaces can be created by application components. Each node maintains its own set of objectplaces, each of which can be given a name unique on the node. An objectplace can be accessed by other application components using its name. This is summarized in Fig. 5.2.

An objectplace by itself is not accessible from nodes other than the node on which it is created. Instead, views and coordination roles are used as a structured way to access and manipulate objectplaces on remote nodes.

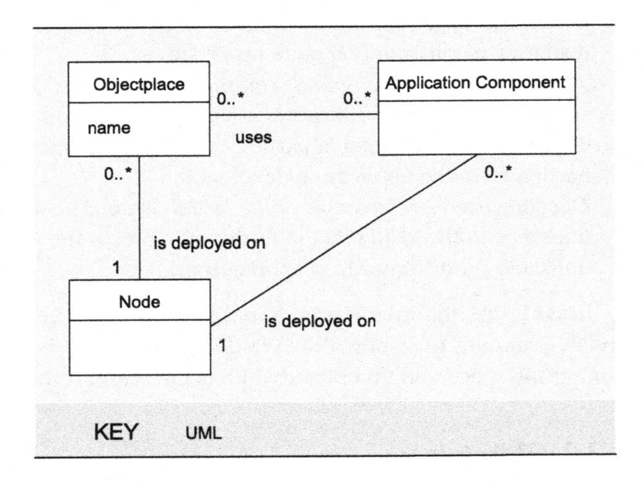

Fig. 5.2 Conceptual model of objectplaces

5.2.2.2 Basic Operations

The three basic operations of an objectplace are `put`, `take`, and `watch`. These three operations add objects to, remove objects from, and observe objects in the objectplace, respectively. All three operations are asynchronous: an application component that executes an operation does not wait for the result, but gives a *callback* as a parameter. When the objectplace has processed the operation, it returns the result of the operation to the callback. Multiple results may be returned over time.

An objectplace is thread-safe: multiple concurrent application components can use the same objectplace safely.

In more detail, the three basic operations on an objectplace are represented as the following methods:

- `put(Collection, Callback)` adds the given collection of objects to the objectplace. When finished, the value true is returned to the callback if all objects were successfully added and false otherwise.
- `take(ObjectTemplate, Callback)` removes the objects matching with the template from the objectplace and returns the matching objects to the callback.
- `watch(ObjectTemplate, EventTemplate, Lease, Callback)` observes the content of the objectplace and returns copies of objects matching the object template to the callback according to the given event template.

An object template is a function that takes a set of objects and returns a boolean value. An object for which the object template returns true is said to *match* with the object template. For the watch operation, application components can select which events are returned using an event template. An event template is a function from the set of possible events to a boolean value. Supported events are `isPresent`, `isPut`, `isTaken`. The watch returns all events for which the event template given by the caller returns true, i.e., a sequence of $\langle event, collection \rangle$ pairs are returned to the callback, where *event* is one of the supported events and *collection* is a collection of objects. A `Lease` serves to unregister watch operations. An application component uses the lease to discard the watch for which the lease was given as argument.

In addition to the basic operations, an objectplace offers one extra operation, `executeAtomically`, to allow the execution of a series of basic operations atomically. For a discussion of this composed operation we refer to [146].

5.2.3 Views

In this section, we describe *views*, the first abstraction supported by the ObjectPlaces middleware. Views enable coordination of application components based on information exchange. Application components declaratively specify in which information they are interested. The middleware builds a view by collecting the required information from objectplaces on remote nodes and maintains the information as nodes move in or out of the view and as the information on remote nodes is changed by other application components.

5.2.3.1 Conceptual Model

A view is a collection of objects that are copies of objects in objectplaces on connected nodes in the network. The middleware builds and maintains a view based on a declarative specification given by an application component. The specification

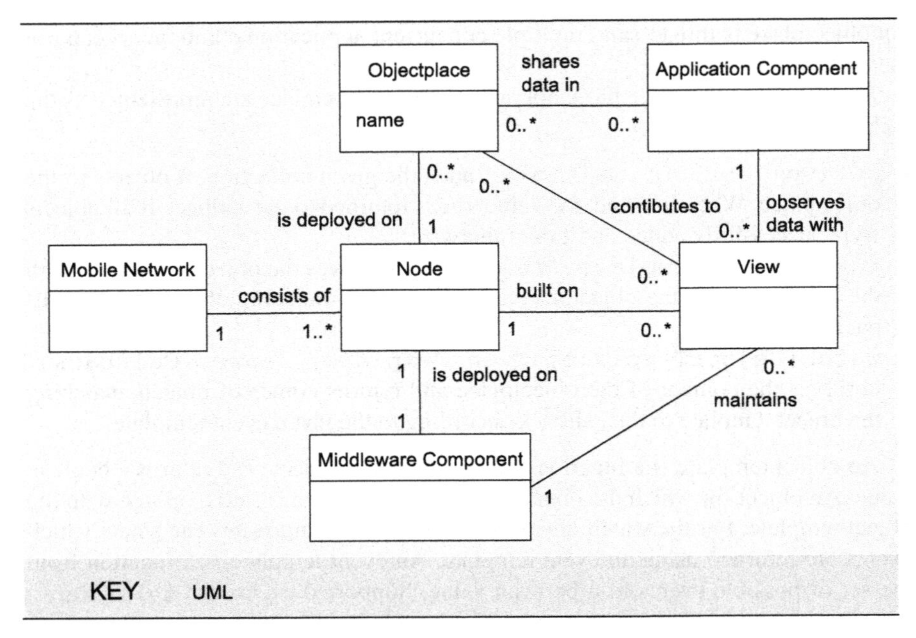

Fig. 5.3 Conceptual model of views

determines the objectplaces that are to be included in the view and the objects that are gathered from those objectplaces. Figure 5.3 shows the concepts and their relations in a conceptual model.

The model shows that application components can use any number of objectplaces to share objects in, and each application component can have any number of different views to observe objects with. The objects in a view are gathered from objectplaces from one or more nodes. These objectplaces or nodes contribute to the view.

The applications are decentralized, i.e., they consist of a set of application components that cooperate as peers. Therefore, the middleware is built as a set of decentralized, cooperating middleware components. Each node hosts one middleware component, which is responsible for providing the necessary services to the application components on that node and coordinating with other middleware components in order to guarantee the middleware's functionality.

Views and objectplaces contribute to the realization of the requirements for middleware for mobile applications (see Sect. 5.2.1) as follows:

- *Allowing Context Awareness.* A view is built by the middleware based on an application-specific constraint on the nodes and objects that should be gathered in the view. A view is thus a representation of the information in the network that is of interest to an application component. Views allow application components to select the information they currently need declaratively.
- *Dealing with Dynamics.* By allowing the application to specify the information it wants to gather by means of a constraint on node properties, the view can be

maintained by the middleware. The application does not need to be concerned with managing changes in interaction partners, as views are kept up-to-date by the middleware.

5.2.3.2 View Management

The main access point of the middleware to start and stop a view is the *View Manager*. To request the construction of a view, an application component uses the `startView` operation that is provided by the view manager. The operation requires three parameters:

1. A *node constraint*, which determines from which nodes in the network the objects in the view are gathered.
2. An *objectplace name*, which determines from which objectplace the objects in the view are gathered.
3. An *object template*, which determines which objects are gathered from the objectplaces on the nodes determined by the previous two constraints.

The node constraint determines which nodes are to contribute to the view based on node properties. Node properties are application-specific properties of a network node, e.g., a node's position. More precisely, a node constraint is a function that takes as arguments the current values of the properties of the viewing node and the current values of the properties of a candidate viewed node, and it returns true if the candidate viewed node should contribute to the view, and false otherwise. The two arguments enable the expression of constraints relative to the viewing node. For example, a view on all nodes within a certain distance from the viewing node needs a node constraint that is a function of both the viewing node's position and the other node's position.

Given the parameters of a view request, the middleware searches the network for nodes satisfying the node constraint. On these nodes, the objectplace whose name is the same as the given objectplace name is found. If the objectplace exists, the objectplace contributes to the view. If the objectplace does not exist, the node does not contribute any objects to the view. This mechanism implies that the objectplace's names are known to all application components building a view and that objectplaces are present during the lifetime of the application.

If an objectplace is found on the node, a watch operation is executed on the objectplace by the middleware component on the viewed node. The watch operation's event template matches with all events. The results of all these watch operations are events indicating the presence, arrival, or removal of objects in an objectplace. The middleware component on the viewed node sends the events to the middleware component on the viewing node.[2] In this way, the middleware

[2] The *viewing node* is the node on which an application node has requested that a view be built. The *viewed nodes* are the nodes that contribute to the view built on the viewing node.

component on the viewing node can keep the view up-to-date with respect to changes in the content of the viewed objectplaces.

Changes in the viewed nodes are handled by the middleware by managing the watch registrations on the objectplaces in the view. Only objects from objectplaces on nodes that satisfy the node constraint remain in the view. When a node moves out of the view, the watch operation on the objectplace of that node is unregistered, and the viewing node is notified that the node moves out of the view. All of the objects that were sent from that node are removed from the view. When a node moves into the view, a watch on its objectplace is registered and the viewing node is notified of the arrival of the new node in the view. Results from the watch operation are sent and the view is updated.

In order to allow the middleware to build and maintain views based on node properties, an application or the middleware on each node maintains node properties for that node in the middleware. Node properties are name–value pairs and may be the result of a sensor readout on the node, e.g., position or another observable property of the node. The middleware imposes no constraint on the form of the values, so they can range from an integer to a complex XML description.

A view is actively maintained by the middleware until it is stopped by the application component. The view manager provides the `stopView` operation to stop a view.

5.2.3.3 Quality of Views

Two important quality attributes that an application developer needs to know about view building and maintenance are as follows:

- Reliability. A perfectly correct view at all times is impossible: at least a transmission delay needs to be taken into account to send the necessary update information to the viewing node. Reliability determines how well the view reflects the actual contents of the objectplaces contributing to the view.
- Efficiency. There is overhead associated with building a view, both computation and communication overhead. Resources used by the middleware cannot be used by the application. Efficiency determines how much overhead the middleware uses to offer its services.

Improvement of reliability is usually at the expense of efficiency: a more timely view needs more updates and more communication.

There is much variation in the quality of mobile networks. At one end of the scale, unpredictable and unreliable mobile ad hoc networks are connected without any network infrastructure besides each node's own network card. On the other end of the scale, wireless LAN networks are supported by access points to relay and amplify communication signals and achieve higher levels of reliability.

An in-depth discussion of implementation strategies for view building and maintenance is out of the scope of this book. Schelfthout [146] discusses two different implementation strategies for different deployment environments. The first implementation strategy describes a protocol for reliable and higher bandwidth wireless

LANs. The second implementation strategy describes a protocol to form views in unreliable mobile ad hoc networks. For each of the implementations, quantitative statements are discussed for two quality attributes: reliability and efficiency.

5.2.4 Coordination Roles

We now describe *coordination roles*, the second abstraction supported by the ObjectPlaces middleware. Coordination roles support the application developer with the design and implementation of dynamic protocols in mobile networks. A coordination role is an abstraction representing the behavior of a component in a protocol. Coordination roles allow the middleware to take over the management aspects of executing a protocol, i.e., the initial discovery of interaction partners in the network and the detection of changes in interaction partners during execution of the protocol. Such management is a main problem of coordination in mobile networks.

5.2.4.1 Conceptual Model

The concepts related to coordination roles and their relations are presented in Fig. 5.4.

A *coordination role* is an abstraction that encapsulates the behavior of one application component engaging in a *protocol*. A *coordination role instance* is a runtime instance of a coordination role. One coordination role can have many coordination role instances at the same time. When a coordination role instance is executing a protocol on behalf of an application component, the component plays the coordination role.

An *interaction session* is the exchange of a series of messages in a protocol by a *group* of coordination role instances played by distinct application components. An interaction session is always started by one application component that starts to coordination play a role by instantiating the coordination role. A coordination role instance that starts an interaction session is called an *initiator*. Coordination roles played by components in the interaction session that participate in an interaction session started by an initiator are called *participants*.

5.2.4.2 Interaction Setup and Maintenance

For the middleware a coordination role is a black box, a unit of behavior that is played by an application component when it is involved in an interaction session. The middleware supports the setup of interaction sessions and the maintenance of the group of coordination role instances in the interaction session as node properties change.

The main access point of the middleware to start and stop interaction sessions is the *Role Activator*. To start an interaction session, an application component can

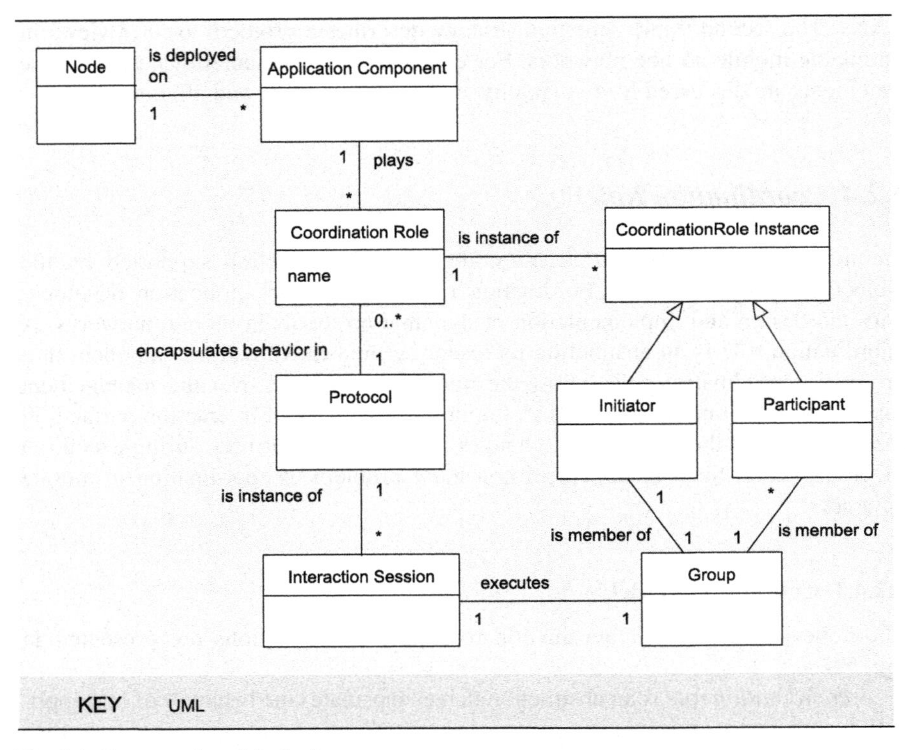

Fig. 5.4 Conceptual model of roles

use the `startInteraction` operation that is provided by the role activator. This operation requires three parameters:

1. An initiator role, which is instantiated by the application component.
2. A node constraint, which specifies the group of nodes on which participants have to be instantiated.
3. The name of a participant role, which will be instantiated on the nodes that satisfy the node constraint.

The node constraint allows constraints based on the individual properties of each participant node and constraints based on relations between properties of initiator node and participant nodes.[3] The former enables the specification of a constraint that compares an arbitrary combination of node property values from a single participant node with a constant value, e.g., to select participant nodes based on their status. The

[3] The *initiator node* is the node on which an application node starts an interaction session using an initiator role. The *participant nodes* are the nodes on which a participant role is activated that participates in the interaction session.

latter enables typical constraints based on differences between properties of nodes, e.g., to select nodes based on distance between initiator and participant nodes.

For all nodes that satisfy the node constraint, the participant role with the given name is instantiated, but only if there are application components on that node that are capable of playing the participant role. To that end, an application component should register the names of the participant roles that it is capable of playing on the node on which the application component is deployed. Whenever a coordination role enters the interaction session, initiator and participants are notified. When a participant enters the interaction session, an asynchronous communication channel is opened between initiator and participant, so that the protocol can be executed.

When a coordination role is instantiated (initiator or participant), the middleware generates a unique identifier that can be used to refer to that role. The initiator and the participants use the `receive` operation to receive messages from a coordination role in the interaction session. To send a message to a participant an initiator uses the `sendToParticipant` operation. The middleware uses the role identifier of the participant to deliver the message. To send a message to all participants, the initiator uses the `sendToParticipants` operation. Participants can send a message to the initiator using `sendToInitiator`.

To support this continuous change in interaction partners, the middleware continuously monitors the node properties and maintains the instantiation of participant roles on the appropriate nodes. Two events can occur in a group of role instances engaged in an ongoing interaction session. First, the properties of a node that is not in the group change, and as a result its node properties satisfy the node constraint. A new participant role is instantiated on the node, and the initiator of the interaction session is notified. The initiator can then take the necessary actions to incorporate the new participant in the interaction session. Second, the properties of a node that is in the group change, and as a result its node properties no longer satisfy the node constraint. The initiator of the interaction session is notified that a participant will be removed. The participant on the node to be removed is notified, so it can clean up. Then, the participant is removed from the interaction session by the middleware. Evidently, only protocols which are able to deal with addition or removal of interaction partners are supported.

In order to allow the middleware to set up a group of coordination role instances based on node properties, the application or middleware on each node maintains node properties for that node in the middleware. Similar to views, node properties are name–value pairs and typically the result of a sensor readout on the node.

The maintenance process continues until the interaction session is closed by the initiator. The role activator provides the `stopInteraction` operation to close the interaction session.

5.2.4.3 Group Membership Guarantees

Regarding the setup and maintenance of interaction sessions, the arrival and removal of a participant in a group are notified to the initiator with a best-effort guarantee.

The update frequency of the node properties and the delay imposed by the underlying communication medium determine the granularity of group updates.

The application can control the update frequency of node properties, taking into account that more updates are likely to cause more overhead. Since the middleware guarantees group updates with the same frequency (i.e., the middleware handles every update to node properties), the application can choose the update frequency such that application requirements are met.

For example, if on every node, the node's position is updated every second, node constraints based on position are updated about every second as well (taking into account jitter on communication delay). In case of mobile nodes, based on the maximum speed, an upper bound can be calculated on the distance a node can travel between two updates. This upper bound can be used to calculate the bounds of the area in which a node is located; this in turn may be important at the application level, e.g., for collision avoidance.

In case a node failure occurs for some reason (hardware or software fault, battery down, etc.), a node is no longer able to communicate. Such failures are in general difficult to handle. In mobile networks with a reliable infrastructure, i.e., with access points, it can be assumed that communication is reliable. A failure detector can then be put in place in order to detect if a particular node cannot be reached anymore. Such a failure can then be relayed to initiators that are in an interaction session with the failed node, as a specific failure event. In this case, the initiator can thus distinguish between a node simply moving out of range and a node that fails. Typically, a failing node requires special measures in a protocol than nodes that move out of range. For example, a protocol that needs to avoid collisions between moving vehicles needs to know whether a vehicle has moved out of collision range or has failed and is still standing approximately at its last known location.

5.3 Middleware Architecture

In this section, we give an overview of the software architecture of the ObjectPlaces middleware. We present the high-level module decomposition of the middleware. Next, we explain group formation, the basis module of the middleware. Then we zoom in on view management and role activation.

5.3.1 High-Level Module Decomposition

Figure 5.5 shows the high-level module uses view of the ObjectPlaces middleware situated in its context.

We summarize the responsibilities of the different modules in turn.

Group formation is the backbone of the ObjectPlaces middleware, providing support for (1) the discovery of groups of nodes that satisfy a node constraint and (2)

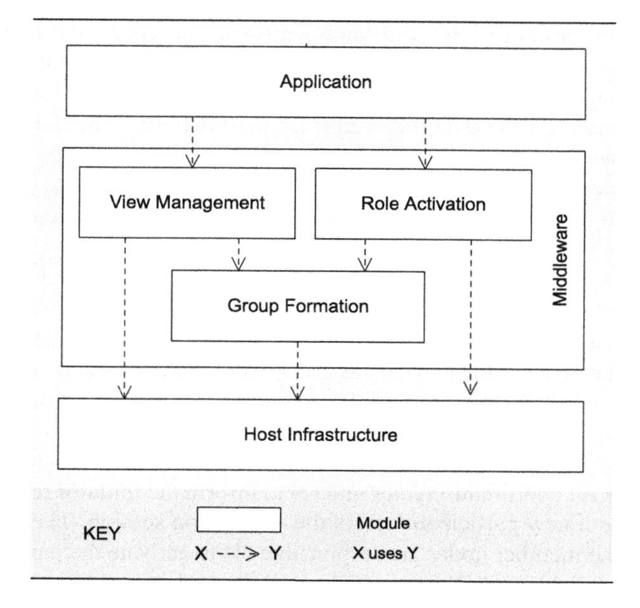

Fig. 5.5 High-level module uses view of the middleware

the maintenance of this group in the face of changes in the properties of the nodes in the network.

Group formation modules on the various nodes use their local set of node properties to determine how the group is formed based on a node constraint. For example, if a group needs to be formed using a distance constraint, the set of node properties on each node contains the node's current position. Node properties are maintained in an objectplace.

The group formation module supports star-formed group formations, where a single leader node forms a group with multiple members. The leader can communicate with all the members of the group, and the members can communicate with the leader. Consequently, the leader is notified of changes in membership of the group, and the members are notified when they join or leave the group. Any number of groups can be formed, and a node may participate in any number of groups simultaneously, both as a leader and as a member.

Star-formed group formation supports both view construction and coordination role activation. For view construction, a view is requested on one node and gathers data is gathered from a number of other nodes in the network. For coordination role activation, an initiator coordination role is activated on one node and interacts with participant coordination roles on other nodes in the network. In both cases, a group is started on one specific node at the initiative of the application, and there are multiple other nodes in the network that need to be part of the group. In neither of the two cases participants need to communicate among each other.

View management provides the service for building views. To build a view, an application component specifies a node constraint and an object template to determine which objects are gathered for the view. The view manager uses the group formation to form and maintain the group. The view manager on the leader node is responsible (1) to send the necessary information to the view managers on the member nodes and (2) to build and maintain a view for the application component based on the data received from the members. The view manager makes the view available for application components in an objectplace. The view managers on the member nodes are responsible (1) to collect matching objects for the view and (2) to notify the view manager on the leader node whenever the situation with respect to the view on the member node changes.

Role activation provides the service for a protocol-based interaction. To start an interaction session, the application component supplies an initiator coordination role that the application component will play in the interaction, a node constraint, and the name of the participant coordination role to be activated. Role activation uses the group formation to find the nodes belonging to the group and keep informed about changes to membership. The role activation module on the node that started the session is responsible (1) to contact the member nodes to activate the desired participant coordination roles and (2) to inform the initiator role when a participant leaves or a new participant enters the interaction session. The role activation modules on the member nodes are responsible (1) to activate the participant coordination role if available and (2) to notify the initiator node when a participant leaves the interaction session or a new participant enters the interaction session.

Design Rationale

The main functional requirements of the ObjectPlaces middleware are the management of views and the management of coordination roles of interaction sessions. With respect to view management, the middleware must be able to build and maintain views in the face of network dynamics based on a node constraint and additional data such as an object template. With respect to management of coordination roles of interaction sessions, the middleware must be able to activate and deactivate coordination roles on the appropriate nodes in the network based on a node constraint and the name of the coordination role that should be activated. These two requirements show that the problem common to both is the resolution of a node constraint to a group of nodes whose properties satisfy the node constraint. This functionality is provided by group formation which is responsible for forming and maintaining a group of interacting nodes. Each of the basic functionalities is encapsulated in a module providing separation of concerns.

Besides functional attributes, the quality of group formation is the major influencing factor on the overall quality of view management and role activation. Important qualities of group formation are reliability and performance. Reliability is a measure of the guarantees that can be accomplished with group formation. Reliability measures how up-to-date the group is and how fast the group is changed in response to changes in the network and node properties. Performance measures the overhead associated with group formation, in particular communication overhead. The main influencing factors on the quality of group formation are the characteristics of the network. Chapter 7 zooms in on the efficiency and bandwidth usage of the middleware for the AGV transportation system.

5.3.2 Group Formation

The group formation module is the backbone of the middleware, providing

1. The discovery of a group of nodes in the network that satisfy a node constraint.
2. The maintenance of this group in the face of changes in the properties of the nodes in the network.

As explained above, the group formation module supports star-formed group formation, see Fig. 5.6.

A single *leader* node (the viewing node or the initiator node) forms a group with many *members* (the viewed nodes or the participant nodes). The leader can communicate with all the members of the group, and the members can communicate with the leader. Consequently, the leader is notified of changes in membership of the group, and the members are notified when they join or leave the group.

To explain the working of group formation (and view management and role activation in the following sections) we use a communicating processes diagram. Communicating processes show a system, or a part of a system, as a set of concurrently executing units and their interactions. The elements of the communicating processes diagram are concurrent units, repositories, and connectors. Concurrent units are an abstraction for more concrete software elements such as task, process, and thread.

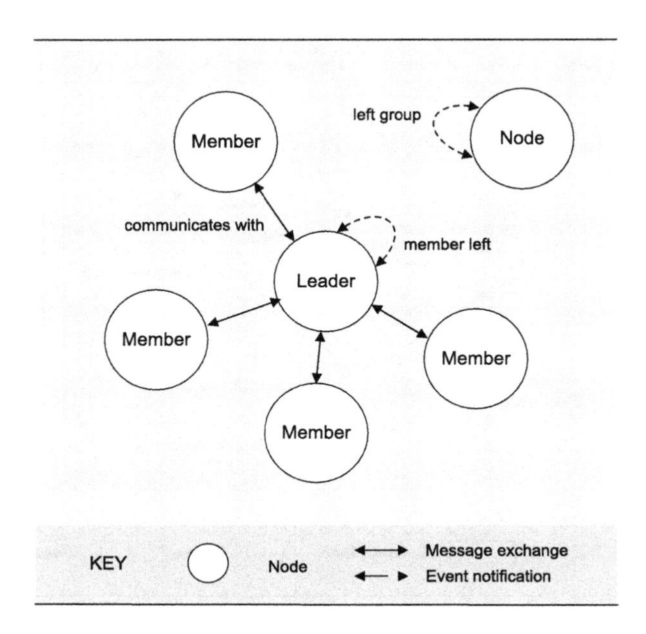

Fig. 5.6 Schematic representation of a group. The leader communicates with the members. On the *right-hand side*: the leader receives a notification of a node that left the group

Repositories are abstractions of more concrete elements such as a buffer. Connectors enable data exchange between concurrent units and control of concurrent units such as start, stop, synchronization.

Figure 5.7 shows the group formation processes in connection with view management and role activation.

We explain the subsequent steps in setting up a group. The number of each step corresponds to the numbers in Fig. 5.7:

1. An application component starts a view or starts an interaction session by sending a request to the View Manager or the Role Activator, respectively.
2. The View Manager and Role Activator process delegate group formation to the Group Formation process (specific actions related to view setup and role activation are explained in detail below). At this point, the node becomes the leader of a new group.
3. The Group Formation process communicates with other nodes using the Message Handler, to determine which nodes are to become a member of the group. The Group Formation process keeps monitoring the group for changes in membership.

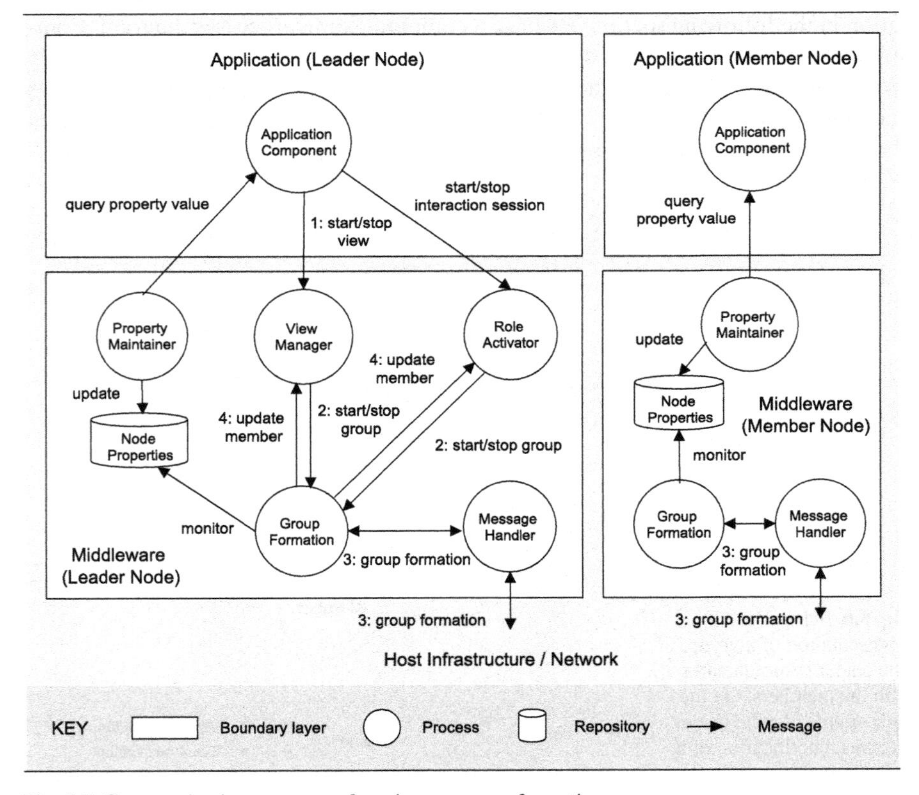

Fig. 5.7 Communicating processes focusing on group formation

4. The Group Formation process notifies the View Manager or Role Activator of the group members and afterward of any changes in group members.

A Property Maintainer process keeps the Node Properties repository up-to-date, e.g., by reading out sensor values. Only property values of the node itself need to be maintained. The Group Formation processes on the various nodes use their Node Properties repository to determine which nodes satisfy a node constraint. The Node Properties repository contains the updated values of all node properties used by the application. For example, if a group needs to be formed using a distance constraint, the Node Properties repository on each node contains the node's current position.

5.3.3 View Management

To build a view, an application component specifies a node constraint, an objectplace name, and an object template. A first step in the construction of a view consists of the resolution of the node constraint to the group of nodes that satisfy the constraint.

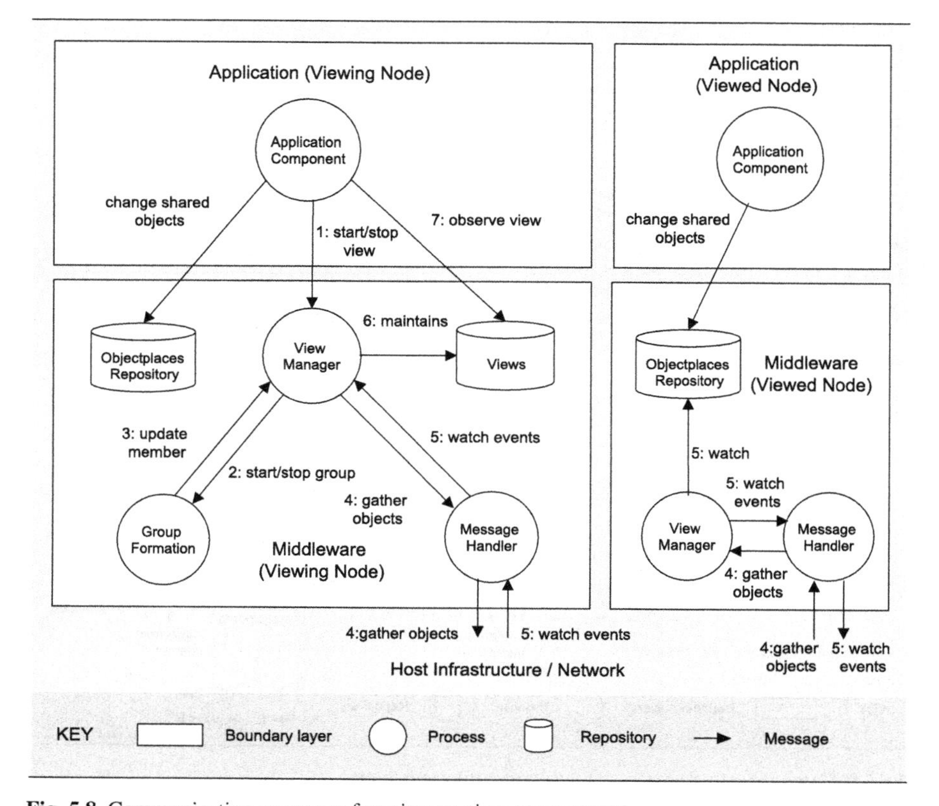

Fig. 5.8 Communicating processes focusing on view management

This task is handled by the Group Formation process. Then, the members of the group are contacted to gather the objects that each of the members contributes to the view.

Figure 5.8 shows the main processes involved in view management. Elements that are not directly relevant for view management, such as the maintenance of node properties, are omitted.

We explain the steps that occur when constructing a view:

1. The View Manager receives a request to build a view from an application component. The request contains the node constraint, objectplace name, and object template.
2. The View Manager passes on the node constraint to the Group Formation process.
3. The Group Formation process forms the group and keeps the View Manager up-to-date with respect to membership (the node on which the view is built becomes the leader of the group).
4. Using the group member information, the View Manager on the leader node sends the objectplace name and object template to the View Managers on member nodes.

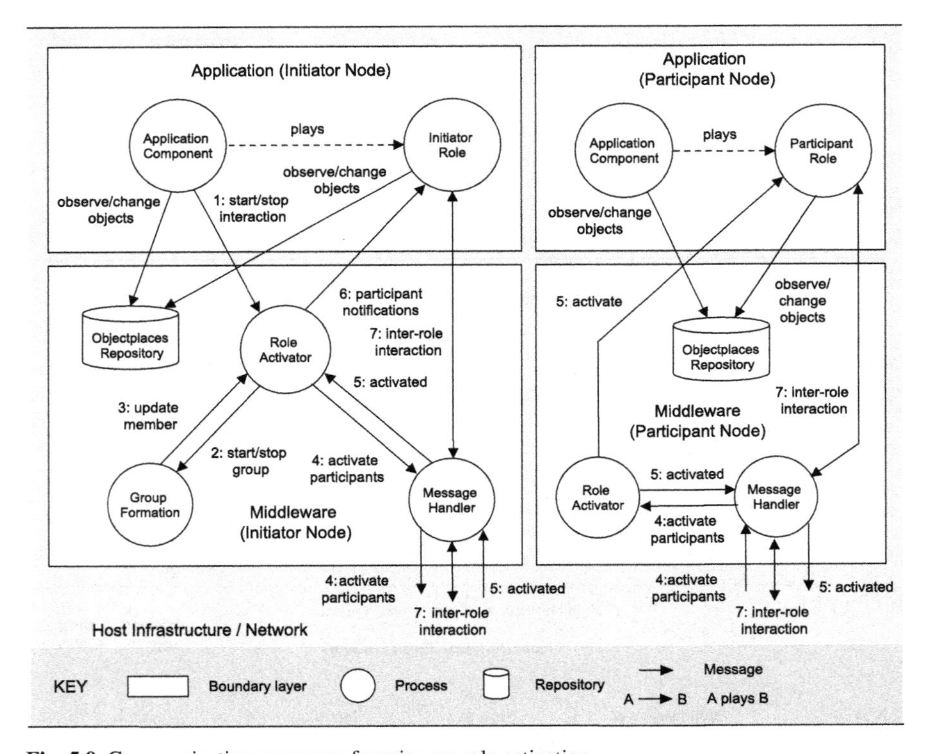

Fig. 5.9 Communicating processes focusing on role activation

5. The member View Managers that are contacted by the leader perform a watch operation on the objectplace with the given objectplace name (shown as Objectplaces Repository), if it exists. If not, no objects from the member node contribute to the view. The events resulting from the watch operation are sent to the leader View Manager.
6. The leader View Manager builds and maintains a view in the Views repository, based on the events received from the members.
7. The application component that requested the view can now observe it in the Views Repository.

Views are maintained continuously. If the View Manager is notified of a change in membership by the Group Formation service, the View Manager updates the view accordingly. Similarly, if View Managers on a member node receive events from a watch, the events are sent to the leader View Manager.

5.3.4 Role Activation

To start an interaction session, an application component supplies an initiator role that the application component will play in the interaction, a node constraint, and the name of the participant role to be activated. A first step in role activation consists of group formation, after which the members of the group are contacted to activate the appropriate role. If the role is activated, a communication channel is set up between initiator and participant roles.

Figure 5.9 shows the main processes involved in role activation.

We explain the main steps that occur during role activation.

1. An application component sends a request to start an interaction session to the Role Activator, specifying an initiator role, a node constraint, and the name of a participant role.
2. The Role Activator gives the node constraint to the Group Formation service.
3. The Group Formation service finds the nodes belonging to the group and keeps the Role Activator up-to-date with respect to membership. The node on which the interaction session is started becomes the leader of the group.
4. The Role Activator on the leader node contacts the Role Activator processes on the member nodes to activate the desired participant role.
5. The member Role Activators activate the participant role, if the role is deployed on the node. The member Role Activator confirms activation of the role to the leader Role Activator.
6. For each participant that is activated, the Role Activator notifies the initiator of the newly activated participant. A handler to the participant that allows the initiator to communicate with the participant is given.
7. The protocol between initiator and participants commences.

The Objectplaces Repository contains the objectplaces that coordination roles and application components on the same node use to coordinate. Application components observe the results or influence the course of the interaction protocol using these objectplaces, and coordination roles use the information in the objectplaces to determine the responses they send.

5.4 Collision Avoidance in the AGV Transportation System

We now demonstrate how views and coordination roles have supported the design and development of the multi-agent system for the AGV application. In this section, we focus on the coordination problem of collision avoidance. We assume that other functionalities such as task assignment, routing, deadlock avoidance, and battery recharging are available. A detailed discussion of task assignment supported with views follows in the next chapter. A solution for deadlock avoidance supported with coordination roles is presented in [146].

5.4.1 Collision Avoidance

Collision avoidance for AGVs is a coordination problem that resembles a *mutual exclusion* problem. Mutual exclusion algorithms are used in concurrent and distributed programs to ensure that several processes do not concurrently use unshareable resources. The un-shareable resources are called *critical sections*. For AGV collision avoidance, critical sections are physical areas on the factory floor that cannot be driven over by several AGVs at the same time. The important difference with classical mutual exclusion problems and collision avoidance is that areas are continuous, so the critical sections are continuous and determined dynamically at run time. In traditional mutual exclusion problems, critical sections are determined at design time and are discrete, i.e., there is a fixed and known number of critical sections that need to be guarded.

While the problem of collision avoidance can be made discrete, for example, by using segments as critical sections (of which there are a known, discrete number on the layout), this solution is not satisfactory, since it does not account for the case where two AGVs need to cross each other on closely located segments. In particular, there may be different types of AGVs working together on the same floor, so two small AGVs may be able to cross at the same time, while two broad AGVs cannot. If maximal flexibility is desired, the best option is then to allow AGVs to describe exactly which area they intend to cross and reserve that area for the AGV, instead of relying on imprecise, worst case discrete critical sections.

As a result, well-known distributed mutual exclusion protocols [136, 155, 101] are not directly usable for AGV collision avoidance. However, the similarities between both problems are still greater than the differences. Consequently, the protocol presented below is a variant on a classical mutual exclusion protocol described by [136].

Research in AGV control systems has tackled the collision avoidance problem [135, 111]. In all approaches, however, collision avoidance is handled together with routing and deadlock avoidance, i.e., an integrated approach to move AGVs from an arbitrary starting point to an arbitrary end point, taking into account the routes and destination of *all* other AGVs on the floor. Because all this information is needed, the approaches are all implemented in a centralized way, i.e., one server calculates all routes for each AGV. Since we study a decentralized architecture, these approaches do not fit our problem.

We have developed a decentralized approach for AGV collision avoidance. The underlying protocol allows decentralized mutual exclusion for continuous critical sections and can be applied to other similar mutual exclusion problems that require fine-grained critical sections.

5.4.2 Collision Avoidance Protocol

To explain the decentralized approach for AGV collision avoidance, we first focus on how the AGV agents avoid collisions without being aware of the underlying collision avoidance protocol. Then, we explain the work behind the scene, i.e., the decentralized mutual exclusion protocol executed by the local virtual environments supported by the ObjectPlaces middleware.

5.4.2.1 AGV Agent Exploits the Local Virtual Environment

In order to drive collision free, an AGV agent exploits the local virtual environment, taking the following actions:

1. The AGV agent determines the trajectory it intends to follow over the layout. The trajectory is determined by *Lock Ahead Distance* parameter that ensures that the AGV moves smoothly and stops safely.
2. The AGV agent calculates exactly which area it is going to occupy on the floor if it drives over its intended trajectory. This area is determined by an AGV's *hull projection*, see Fig. 5.10. A *hull* is the physical area an AGV occupies on the floor. A hull projection is the union of a set of hulls, projected along the AGV's intended path in small increments. The hull projection determines accurately the space an AGV will occupy if it would drive over the path; so, if a number of hull projections of a set of AGVs overlap, the AGVs are on collision course.
3. To avoid collisions, an AGV agent tries to reserve the area represented by the hull projection for exclusive use. Therefore the agent marks the path it intends to drive in the local virtual environment[4] with a *requested hull projection*. This projection contains the agent's identity and a priority that depends on the transport the AGV is handling.

[4] For convenience, we use "local virtual environment" to refer AGV local virtual environment in the remainder of this chapter.

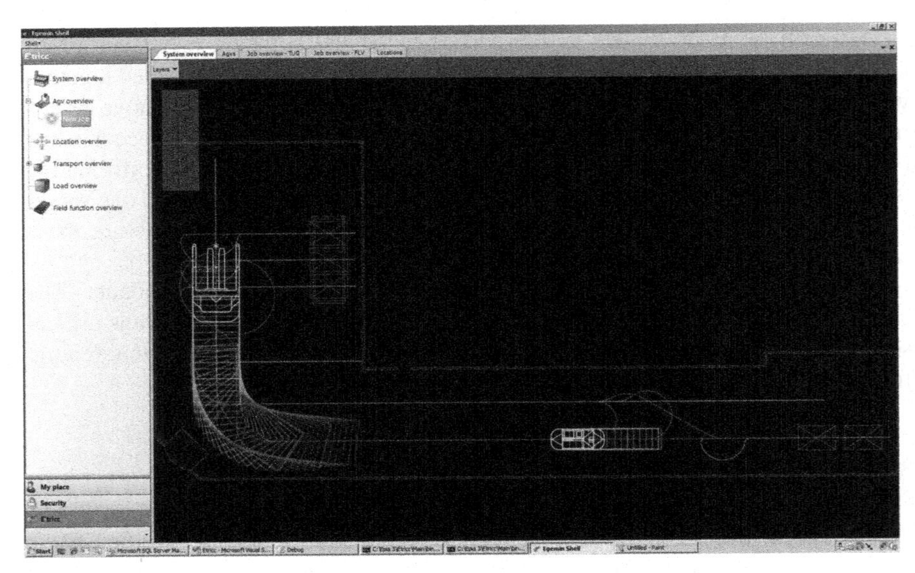

Fig. 5.10 A top-down view of a factory floor with two AGVs which are projecting a hull

4. The agent perceives the local virtual environment to observe the result of its action.
5. The agent examines the perceived result. There are two possibilities:

   a. The requested hull projection is marked as a *locked hull projection*: it is safe to drive.
   b. The hull is not marked as locked: this means that the agent's hull projection conflicted with that of another AGV agent. The agent may not pass; at this point the agent may decide to wait and look again at a later time or remove its requested hull projection and take another path altogether.

Since the AGV steering system, E'nsor, must be instructed to drive segment per segment (i.e., the level of granularity is one segment), an AGV's requested hull projection spans at least one segment. When an AGV is driving, the AGV agent releases the parts of its locked hull projection behind it, so that other AGVs may pass. Note that AGVs cannot completely clear their locked hull projections, since an AGV at least needs to keep a lock on the area it is currently standing.

5.4.2.2 Decentralized Mutual Exclusion Protocol

We now shift our focus to the AGV's local virtual environment which must resolve conflicts with the local virtual environments of other AGVs that intend to move and make sure that the requested hull projection becomes locked eventually. To this end, the local virtual environment of the AGV agent that requests a new hull projection executes a mutual exclusion protocol with local virtual environments of nearby AGVs.

In order to guarantee safety and save bandwidth, the subset of local virtual environments with which a requesting local virtual environment interacts must include the local virtual environments of all AGVs with which the AGV of the requesting local virtual environment might collide. Figure 5.11 illustrates how safe subset of AGV local virtual environments is determined.

A requesting local virtual environment interacts with other local virtual environments whose *hull projection circle* overlaps with the hull projection circle of requesting local virtual environment. The hull projection circle is defined by a center point, which is the position of the AGV itself, and a radius, which is equal to the distance between the AGV and the furthest point on its hull projection. So, overlapping circles indicate to a first approximation that two AGVs are within collision range. This approximation has the benefit that it narrows down the possible candidates for interaction significantly, while each AGV only needs limited knowledge about other AGVs to determine interaction partners (i.e., position and hull radius).

Due to the mobility of the AGVs, a new AGV entering collision range should be taken into account when executing the collision avoidance protocol, and an AGV leaving collision range can be disregarded. Using the middleware support, the collision avoidance protocol is modeled by two roles: a Requester and a Voter role. To lock a new hull projection, the local virtual environment activates a Requester role, asking the activation of Voter roles with a node constraint that selects all AGVs within collision range:

$$c_{\text{node}}(V_{\text{init}}, V_{\text{part}}) = dist(V_{\text{init}}.pos, V_{\text{part}}.pos) \leq V_{\text{init}}.rhull + V_{\text{part}}.rhull$$

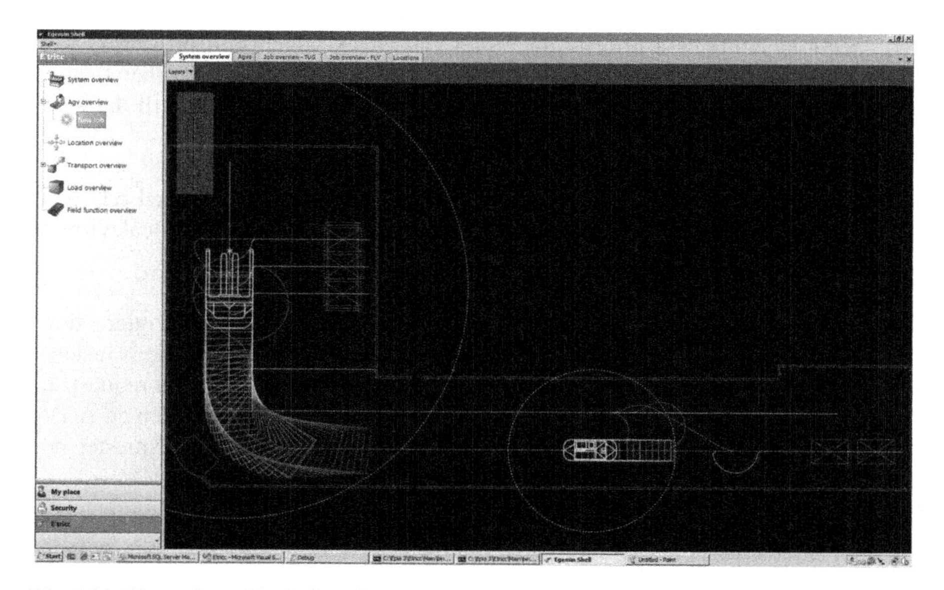

Fig. 5.11 Illustration of the hull projection circle

V. pos denotes the current (*x*,*y*) position, *rhull* the current hull radius. From this constraint, it is clear that the middleware needs the AGV's positions and current hull radii to determine where voters should be activated, so the application updates this information in the node properties repository. On each AGV, the AGV's position and hull radius are updated every second. The middleware takes care of disseminating positions and hull radii to other AGVs. So, a small amount of data is sent to all AGVs, in order to allow the AGVs to execute the collision avoidance protocol in smaller groups. To instantiate the necessary Voter roles, the middleware finds all the AGVs in the system whose properties satisfy the node constraint. The Requester role is notified of these Voter role instances, after which the collision avoidance protocol can be executed.

Once the group is settled, to lock a requested hull projection, the local virtual environment executes the following mutual exclusion protocol with the local virtual environments in collision range:

1. The requester sends a `Request(HullProjection)` message to voters.
2. The voters check whether the requester's hull projection overlaps with their hull projection. There are three possibilities for each of the requested voters:

   a. No hull projections overlap. The voter sends an `allow` message to the requester.
   b. The requester's hull projection overlaps with the voter's hull projection, and the voter's hull projection is already locked. The voter defers to send an `allow` message until the lock on the overlapping area is released.
   c. The requester's hull projection overlaps with the voter's hull projection, and the voter's hull projection is not locked. Since each of the requested hull projections contains a priority, the voter can check which hull projection has precedence. If the requester's hull projection has a higher priority than that of the voter, the voter replies `allow`; otherwise the voter defers until the lock on the overlapping area is released.

3. The requester waits for all votes to come in. If all voters have voted `allow`, the requested hull projection can be locked and the state of the local virtual environment is updated.

When a new AGV enters collision range while a collision avoidance interaction session is in progress, this is detected by the middleware and a Voter role is instantiated on that AGV. The Requester is notified, and in response sends a request to the new Voter, and also waits for the allow message from that AGV. When an AGV moves out of collision range, the Requester is notified, and so the Requester no longer waits for that Voter.

Intuitively, the protocol is *safe*, i.e., collision-free movement is guaranteed, because for each two AGVs with overlapping requested hull projections, exactly one request is allowed. However, a closer examination reveals that two problems must be solved to guarantee safety of the protocol:

1. Group formation may be out of date. The middleware sends update messages to inform AGVs of new positions and hull radius. However, this information is updated once per second, and there is a transmission delay. The information an AGV has about other AGVs thus may not reflect the current situation. As a result, an AGV may not send a request to another AGV that is within collision range and erroneously assume that it is safe to lock a hull projection.
   This problem is solved as follows. Given the update interval of 1s for position and hull radius, and a maximum message delay t_{delay}, every Requester must wait a minimal safe time of 1 s plus t_{delay} before closing a session and locking a hull. This delay ensures that the middleware has had time to exchange the requesting AGV's new position and hull radius with other AGVs, so that each AGV's information is up-to-date with respect to the requesting AGV. In practice, since t_{delay} is much smaller than 1 s, the safe time is set conservatively to 2 s.
2. Due to communication delays, group formation may be temporarily inconsistent. In particular, when two local virtual environments start an interaction session to execute the collision avoidance protocol an error may occur when a voter on an AGV that is also requesting sends an allow message to an AGV that is not in the AGVs group, Fig. 5.12 shows a scenario.

To enforce consistency, we add the condition that a voter role may only allow a request if the requesting AGV is also in the collision avoidance group as seen by the AGV on which the voter is deployed.

Appendix C describes the collision avoidance protocol in detail and provides a proof of safety.

5.4.3 Software Architecture: Communicating Processes for Collision Avoidance

We now illustrate how collision avoidance is dealt with in the software architecture of the AGV transportation system. Figure 5.13 shows the communicating processes diagram for collision avoidance.

The diagram presents the basic components of the AGV control system (AGV Agent, AGV Local Virtual Environment, and Middleware) and overlays them with the main processes and repositories involved in collision avoidance; compare the module decomposition view of the AGV transport system in Fig. 4.12, the collaborating components of the AGV agent in Fig. 4.14, and the collaborating components view of the local virtual environment in Fig. 4.19. We explain the subsequent interactions between the main processes involved in locking a requested hull projection for collision avoidance. The number of each step corresponds to the numbers in Fig. 5.13:

1. The Collision Avoidance process of the AGV agent, which is part of the decision making component, requests the Action Manager process a hull projection.

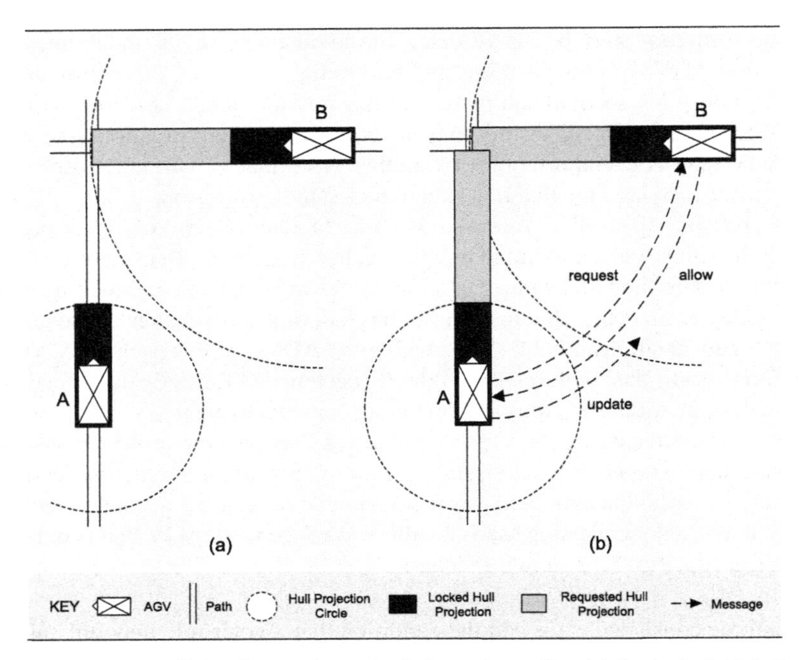

Fig. 5.12 A possible collision if group formation is inconsistent. Part (**a**) shows the initial situation: two AGVs driving toward each other, each with a locked hull projection. The *circles* show the AGV's hull radius. AGV B has a pending requested hull projection but has not requested an allow from AGV A since AGV A is not within collision range. Part (**b**) shows what can happen if AGV A also requests a new hull projection that overlaps with AGV B's requested hull projection. AGV A sends a request to AGV B, which, if AGV A's priority is higher than AGV B's, is allowed by AGV B. AGV A's request message, however, has arrived faster than the update message indicating AGV A's new hull radius to AGV B. At this point AGV B can decide to lock its own requested hull projection, since it is not aware that it should send a request to AGV A. Likewise, AGV A has received an allow vote from AGV B, so it too can lock its requested hull projection. Collision is then imminent

2. Action Manager instantiates the Requester role and the corresponding Requester process. The requester role adds the requested hull to the Collision Avoidance Objectplace.
3. Action Manager requests Role Activator to start the collision avoidance protocol to lock the requested hull projection.
4. Role Activator uses Group Formation to start the group.
5. Group formation communicates with the Group Formation processes on the other AGVs to determine which AGVs are to become a member of the group.
6. Group Formation notifies the Role Activator of the group members, and afterward in case a member leaves the group or a new member joins the group.
7. Role Activator contacts the Role Activators on the member nodes, i.e., the AGVs that are in collision range, to activate the Voter role.

8. The Role Activators on the member nodes inform Role Activator that the Voter role is activated.
9. Role Activator in turn notifies Requester.
10. When all the Voter roles are activated, Requester starts the collision avoidance protocol sending requests to the Voters with the requested hull projection.
11. Each Voter sends an allow message when the requested hull projection does not overlap with their hull projection.
12. As soon as the Requester has received an allow message from all the voters, it locks the hull in the Collision Avoidance Objectplace repository.

Subsequently, the Hull Maintainer process, which is part of the synchronization module of the AGV local virtual environment (see Fig. 4.19), observes the hull change and updates the hull representation in the State repository of the AGV local virtual environment. Finally, the Collision Avoidance process uses the Perception process to sense the status of the hull projection. The Collision Avoidance process notices that the hull is locked and the AGV can move on.

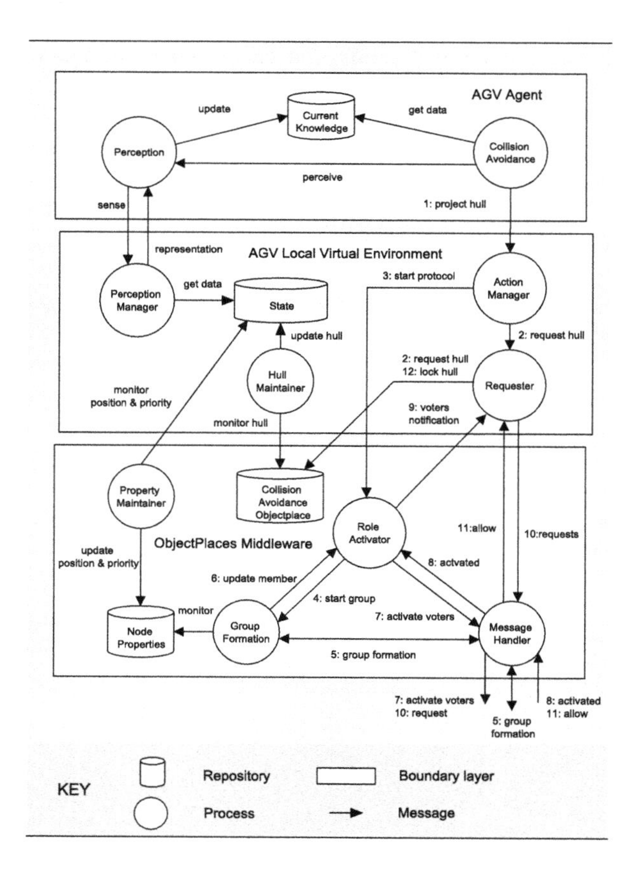

Fig. 5.13 Communicating processes for collision avoidance

5.5 Summary

In this chapter, we discussed the crucial role of middleware support for multi-agent systems. We gave an overview of the role of middleware for supporting the development of distributed systems, and we discussed the multiple layers of middleware in distributed software systems in general. Then we discussed how typical middleware for multi-agent systems maps on the different middleware layers. We explained that the focus of middleware support for multi-agent systems is on the domain-specific middleware layer. Such domain-specific middleware simplifies application development by offering programming abstractions that hide lower level details from the application developer, and it accelerates application development by encapsulating generic, reusable functionalities to support the programming abstractions.

The case study presented ObjectPlaces, a middleware for mobile systems. ObjectPlaces targets applications that are characterized by context awareness, dynamic operating conditions, and decentralization of control. These characteristics closely connect with many applications targeted by multi-agent systems.

We presented the two programming abstractions offered by the middleware: views and coordination roles. A view is a representation of data objects shared by application components in objectplaces on other nodes in the network. A coordination role is an abstraction that encapsulates the behavior of an application component in a protocol. We discussed the software architecture of the middleware and motivated the rationale for the architectural design. We used communicating processes diagrams to precisely describe the internals of group formation, view management, and role activation.

Finally, we explained how we have applied ObjectPlaces in the AGV transportation system. Our particular focus in this chapter was on collision avoidance. To avoid collisions, an AGV agent coordinates with other AGV agents by projecting a requested hull in the local virtual environments that demarcates the area the AGV intends to drive. The local virtual environments of the AGVs in collision range resolve conflicts by executing a mutual exclusion protocol. Dealing with dynamics and context awareness is a difficult problem in the AGV application. By applying coordination roles, we showed that the application developer can abstract from low-level details; the tedious but important tasks, such as finding the AGVs in collision range handling, dealing with AGVs that leave, and new AGVs that enter collision range, are handled by the middleware. The middleware support has shown to be invaluable in the design and development of this real-world multi-agent system.

Chapter 6
Task Assignment

Task assignment in multi-agent systems is a complex coordination problem, in particular in systems that are subject to dynamic and changing operating conditions. To enable agents to deal with dynamism and change, adaptive task assignment approaches are needed. In this chapter, we study two approaches for adaptive task assignment that are characteristic for two classical families of coordination mechanisms for task assignment. In particular, we study and compare a field-based approach for task assignment (FiTA) with a protocol-based approach (DynCNET). In FiTA, tasks emit computational fields in a virtual environment that attract idle agents. Agents follow the gradient of the combined field that guides them toward tasks. DynCNET is an extension of the well-known contract net protocol CNET [151], with "Dyn" referring to support for dynamic task assignment. Both FiTA and DynCNET enable task assignment in the system based on local interaction among agents and allow for adaptation of task assignment during delayed commencement. Yet, the approaches differ in the manner agents realize task assignment. In FiTA, agents use simple rules that guide them toward tasks, providing an emergent solution for task assignment. Contrarily, in DynCNET agents use explicit selection mechanisms and can negotiate about task assignment. Our focus is on systems with homogeneous tasks that can be executed by individual agents. We do not consider complex tasks, for instance composite tasks that have to be divided among agents, or a combination of related tasks that have to be executed by a single agent. This perspective allows us to focus on the basic challenges of task assignment in systems that are subject to dynamic and changing operating conditions.

We use the AGV transportation system as a concrete case to illustrate and validate the two approaches for adaptive task assignment. After a brief description of the traditional approach for task assignment used by Egemin, we introduce the two approaches for adaptive task assignment. Then, we evaluate the approaches based on test results obtained from a simulated industrial AGV transportation system. The evaluation compares (1) the performance of both approaches (throughput and bandwidth usage); (2) a number of important quality attributes, including flexibility (adapt to dynamics that happen during task assignment), openness (take into account agents that enter/leave the system in the process of task assignment), and robustness to message loss (degrade gracefully with increasing loss of messages); and (3) the

D. Weyns, *Architecture-Based Design of Multi-Agent Systems*,
DOI 10.1007/978-3-642-01064-4_6, © Springer-Verlag Berlin Heidelberg 2010

complexity and support to engineer the approaches. The chapter concludes with a summary.

6.1 Schedule-Based Task Assignment

Traditional AGV systems deployed by Egemin use so-called schedule-based task assignment. A schedule defines a number of rules that are associated with a particular location and is only valid for that location. The rules define what an AGV has to do when it visits the schedule's associated location. The AGV transportation system determines when the schedule is triggered depending on the current situation of the system such as the current position and status of the vehicles, loads. Schedule-based task assignment has two important advantages: (1) the behavior of the system is deterministic and (2) task assignment can precisely be tailored to the requirements of the application at hand. Unfortunately, the approach has also disadvantages. First, the approach is complex and labor intensive. Layout engineers have to define all the rules manually. Second, the assignment of transports[1] is statically defined. The approach lacks flexibility. To improve flexibility, dynamic scheduling is introduced. Dynamic scheduling allows reassignment of jobs when an AGV is able to perform more opportune work. Yet, the approach remains limited since it only allows an AGV to perform a new pick job in very specific circumstances, for example, when an AGV drives to a park location or when it performs an opportunity charge action.

Since the execution of a transport requires a preceding effort of an AGV before the transport can actually be executed, we call this characteristic delayed commencement, deferring final task assignment until an AGV that picks a load will benefit the flexibility of the system. The decentralized architecture aims to provide an approach for task assignment that enables AGVs to flexibly switch task assignment when opportunities occur while the AGVs drive toward loads. In the remainder of this chapter, we present FiTA and DynCNET, the two approaches for adaptive task assignment that we developed for the AGV transportation system, and we make a tradeoff analysis. The evaluation is performed on a layout of a real AGV system that is implemented by Egemin and we use standard transport profiles. In the tests, we make abstraction of a number of concerns, such as charging of the batteries of AGVs, calibration of the vehicles, and persistency of data to recover from failures. It is common practice when testing specific properties of AGV transportation systems to focus on the concern under test [70].

6.2 FiTA: Field-Based Task Assignment

The basic idea of field-based task assignment is to let each idle agent follow the gradient of a field that guides it toward a task that has to be executed. The agents continuously reconsider the situation and task assignment is delayed until the execu-

[1] In the context of the AGV application, a *transport* and a *task* are synonyms.

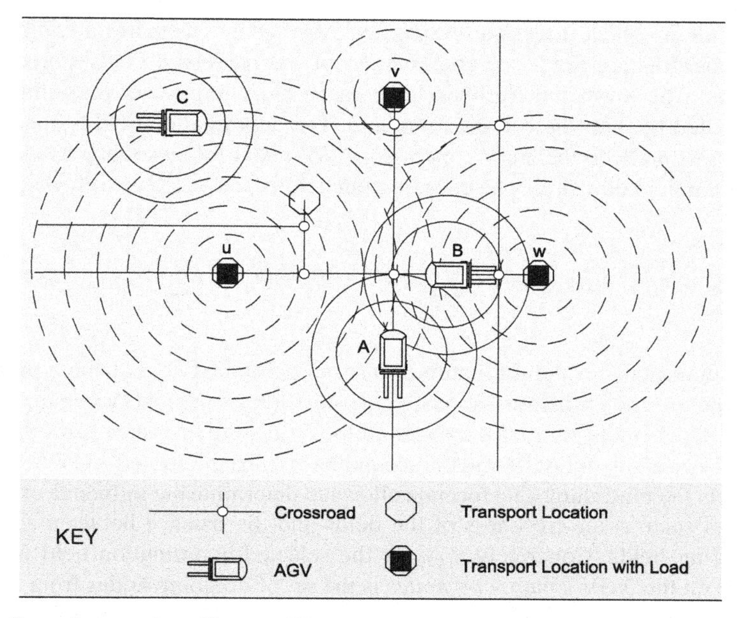

Fig. 6.1 Example scenario to illustrate FiTA

tion of the task starts, which benefits the flexibility of the system. To explain FiTA, we use the scenario shown in Fig. 6.1.

Both AGV agents and transport agents emit fields in the local virtual environment. Transport fields attract idle AGVs. However, to avoid multiple AGVs driving toward the same transport, AGVs emit repulsive fields. AGV agents combine perceived fields and follow the gradient of the combined fields that guide them toward pick locations of transports. Fields have a certain range and contain information about the source agent. The fields of the AGV agents have a fixed range and contain the identity of the AGV and its current location. The range of transport fields is variable and depends on the priority of the tasks. Transport fields contain the identity of the transport, the location of the load, and the actual priority of the transport. Fields are refreshed at regular times, according to a predefined refresh rate. The spreading of the fields is a responsibility of the local virtual environments.

6.2.1 Coordination Fields

When an idle AGV agent perceives fields, it stores the data contained in these fields in a *field cache*. The field cache consists of a number of cache entries. Each cache entry contains the most recent data contained in a field and a *freshness*. The freshness is a measure of how up-to-date the cached data is. For example, in Fig. 6.1 the field cache of AGV A will consist of three entries, one for transport *u*, one for transport *w*, and one for AGV B.

To decide in which direction to drive, an AGV agent calculates a *coordination field*. A coordination field is a combination of the perceived fields stored in the field cache. The lower the freshness of a cache entry, the lower the influence of the associated field on the coordination field. The coordination field is constructed from the next node on the AGV's path. An AGV agent follows the *gradient* of the coordination field downhill. The coordination field is computed as follows:

$$F_{\text{calc}} = min|_{j \,\in\, \text{out\_nodes}} \left(\delta \sum_{i=1}^{n_T} F_{i,j}(1+\phi_i) + (1-\delta) \sum_{k=1}^{n_A} F_{k,j}(1+\phi_k) \right)$$

The formula calculates the minimum of a set of combined fields from a particular node on the warehouse layout. For each possible direction the AGV can move from this node, the formula computes the sum of the fields (the first term sums the transport fields sensed by the AGV and the second term sums the sensed AGV fields) and then selects the minimum. The formula allows to determine the influence of various parameters such as the freshness of the fields and the balance between attracting and repelling fields. Concretely, F_{calc} is the selected coordination field from the next node on the AGV's path. *out_nodes* is the set of outgoing nodes from the next node. n_T is the current number of entries of transport fields in the field cache and n_A the number of entries of AGV fields. δ is a weight coefficient that determines the contribution of transport fields relative to AGV fields. $F_{i,j}$ is the field strength of transport i of the field cache on the next node via node j. ϕ_i is the freshness coefficient of the sensed field of transport i. $F_{k,j}$ is the field strength of AGV k of the field cache on the next node via node j, and ϕ_k is the freshness coefficient of the sensed field of AGV k.

$F_{i,j}$ is computed as follows:

$$F_{i,j} = \frac{Router(l_i,j)}{p_i}$$

$Router(l_i,j)$ calculates the shortest path distance from l_i, the location of transport i, to the next node on the AGV's path via node j. p_i is the actual priority of transport i.

$F_{k,j}$ is computed as follows:

$$F_{k,j} = Router(l_k,j)$$

$Router(l_k,j)$ determines the shortest path distance from l_k, the location of AGV k, to the next node of the AGV via node j.

As an illustration, in the left part of Fig. 6.2, AGV A calculates the coordination field on the node in front. Although transport w is closer, the coordination field will guide AGV A toward transport u. This is the result of the repulsive effect of AGV B. It would have been ineffective for AGV A to drive toward transport w, since AGV B is closer and is maneuvering toward this transport.

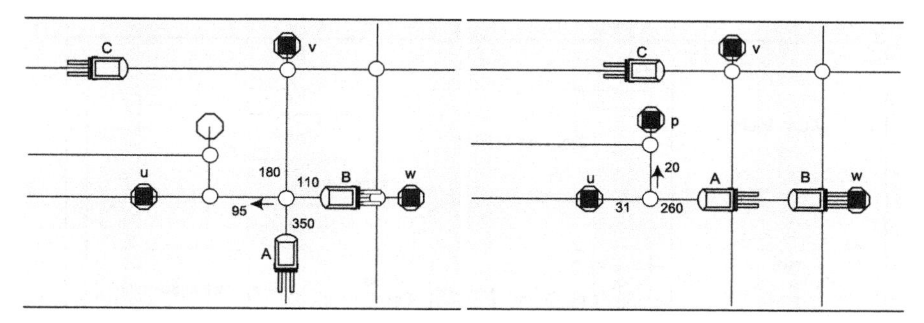

Fig. 6.2 Two successive scenarios in which AGV A follows the gradient of the combined fields. For clarity, we have not drawn the fields. The key is the same as in Fig. 6.1

6.2.2 Adaptive Task Assignment

Final task assignment in FiTA is delayed until an AGV actually reaches a pick location and picks up the load. This allows agents to adapt the assignment of tasks, while the AGVs drive toward loads. By delaying task assignment, FiTA can cope with changing circumstances. An example is shown in the right part of Fig. 6.2 where a new transport suddenly pops up. While AGV A is driving toward transport *u*, a new transport *p* appears close to AGV A. Since no transport has been assigned to AGV A yet, it can drive toward transport *p*.

6.2.3 Software Architecture

Figure 6.3 shows a collaborating components view with the main components of the AGV agent, the AGV local virtual environment, and ObjectPlaces in FiTA. Transport agents have a similar but more simple decision making component as AGV agents since these agents only have to deal with emitting fields.

First, we discuss the various components of the AGV agent that deal with field calculation. Then, we zoom in on the components of the local virtual environment that deal with field management.

Field Cache. This repository stores the information of fields of other AGV agents and transport agents in cache entries.

Router. The router uses a map of the warehouse layout with nodes and segments to calculate paths and distances from one node to another. For testing, we have used a static router that uses the A* algorithm [68]. However, the approach is compatible with a dynamic router that would take into account dynamic runtime information such as traffic distribution.

Field Calculator. The field calculator computes the coordination field from the last selected target node by combining the perceived fields from the field cache. The higher the freshness of a cache entry, the more the influence the field associated with the cache entry will have on the construction of the coordination field. Thus,

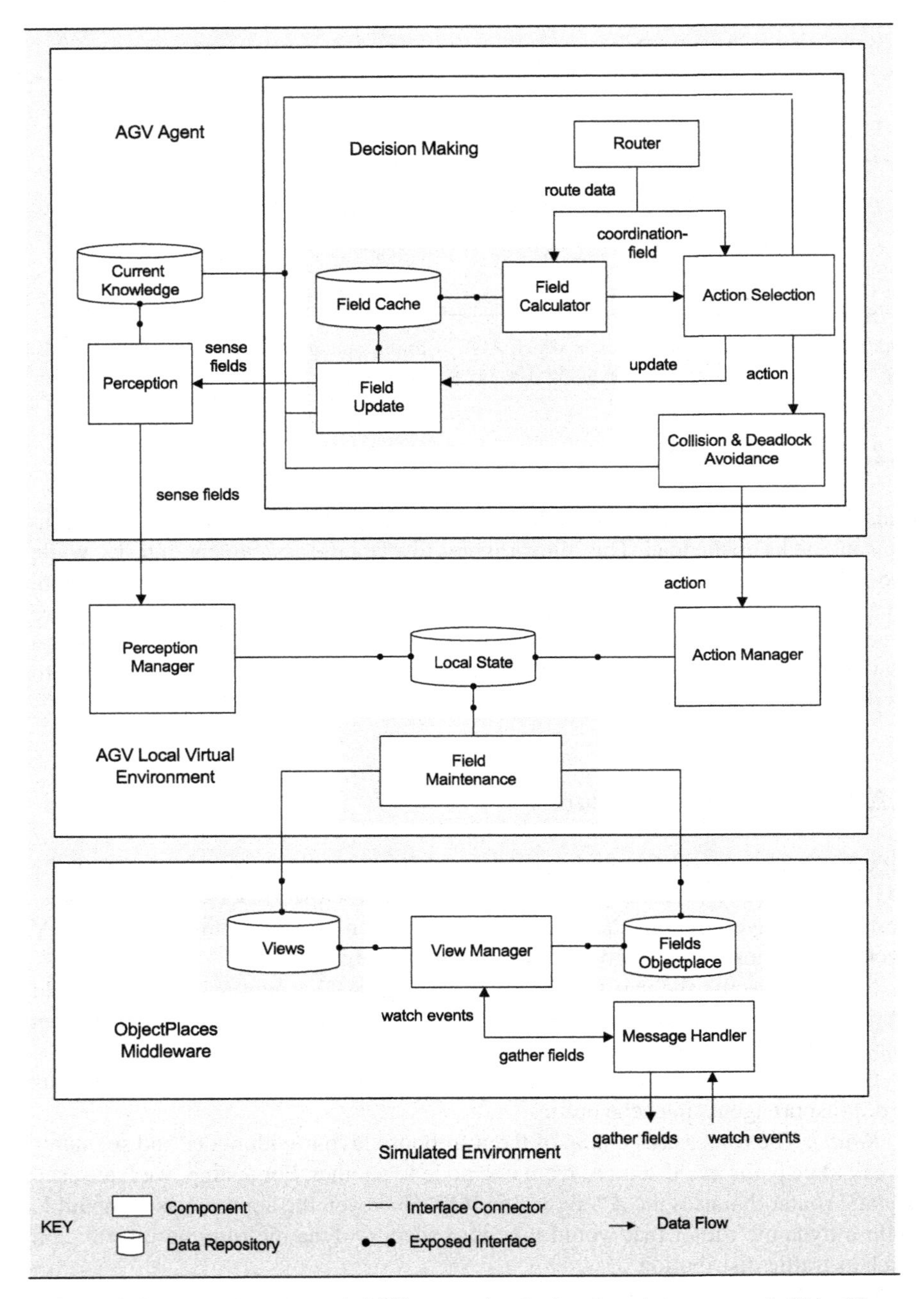

Fig. 6.3 Software components of AGV agent, local virtual environment, and ObjectPlaces involved in FiTA. The elements in the *shaded area* of the local virtual environment and the middleware deal with field management and the elements of the AGV agent deal with field calculation

although still used, less importance is given to outdated information. The field calculator makes use of the router to calculate the values of the coordination field in different directions. The AGV follows the gradient of the coordination field downhill as driving direction.

Field Update. The field update component requests perception updates (via the Perception component) to update the field cache of the AGV agent. Field update requests are periodically invoked by the action selection component.

Action Selection. The action selection component continuously reconsiders the dynamic conditions in the environment and selects appropriate actions to perform the agent's tasks. We illustrate action selection of the AGV agent with a number of example rules:[2]

```
{Action selection rules of AGV agent}
R1: (ready-to-pick) -> {action = pick}
R2: (reserved-path < LookAheadDistance)
      -> { compute coordination-field;
           action = reserve-node }
R3: (ready-to-move) -> {action = move}
```

Rule R1 states that the AGV agent selects a pick action when the AGV is ready to pick a load. Rule R2 states that the AGV agent reserves a next node on its way to a load if the current length of its reserved path is less than the predefined path length LookAheadDistance. Locking the path in advance according to the LockAheadDistance parameter ensures that the AGV moves smoothly and stops safely. The third rule states that the AGV agent selects a move action if it is ready to move on.

Action selection passes the selected action to the Collision & Deadlock Avoidance component. If applicable, this component locks the required path to execute the selected action. As soon as the path is locked, the action is invoked in the AGV local virtual environment. When the AGV has picked up a load, it will inform the transport agent and execute the transport. The following two high-level descriptions summarize the behavior of the agents during task assignment:

```
{Action selection AGV agent}
while idle
 do repeat with constant frequency {
  1. Sense fields and update the field-cache
  2. Select action
  3. Perform action in AGV local virtual environment
}

{Action selection transport agent}
```

[2] The format of the rules is defined as
`(condition)` → `{optional computation; selected action}`

```
while not assigned
 do repeat with constant frequency {
  1. Calculate priority
  2. Update status in the TB local virtual environment
}
```

Now, we zoom in on the components of the AGV local virtual environment and the ObjectPlaces middleware related to field management.

Local State. This repository of the AGV local virtual environment stores the values of fields of AGVs and transports (among other states).

Field Maintenance. The AGV local virtual environment is responsible for spreading the fields. Field maintenance encapsulates a dynamic process that maintains the local fields. It takes into account the status of the local agent such as the position of an AGV and the priorities of transports and the information about AGVs and transports received from other local virtual environments.

Fields Objectplace. In this repository, field maintenance maintains the field of the AGV. At the transport base, the TB local virtual environment maintains the status of the fields of the transport agents in the fields objectplace based on their priority. Fields are removed when the corresponding load is picked.

Views. In this repository, the view manager builds up the view with relevant fields for the local agent. The view manager gathers the fields of neighboring AGVs and transports via the message handler. The view manager maintains the group of involved nodes based on the watch events it receives from the other nodes. The fields in the view repository are monitored by field maintenance that use them to maintain the local state of the AGV local virtual environment.

6.2.4 Dealing with Local Minima

A well-known problem with field-based approaches is the problem of local minima [90]. We explain how FiTA deals with two common causes of local minima: the topology of the layout and the neutralization of fields.

Since AGV vehicles are confined to follow predefined paths in the environment, the problems with local minima caused by the topology of the layout could be avoided relatively easily. Consider the situation on the left in Fig. 6.4 with AGV A and two transports u and v. If the values of the fields would be based on Euclidean distance, AGV A would drive toward transport u; however, it would be trapped in a local minimum at node 1. By making the strength of the field on a particular position proportional to the shortest path distance between this position and the source of the field, local minima are avoided. When applied to the example in Fig. 6.4, since the shortest path distance from AGV A to transport v is much smaller as to transport u, the attracting field of transport v will be much smaller than that of transport u. As such, AGV A will turn right at node 1 (gradient downhill) and drive toward transport v.

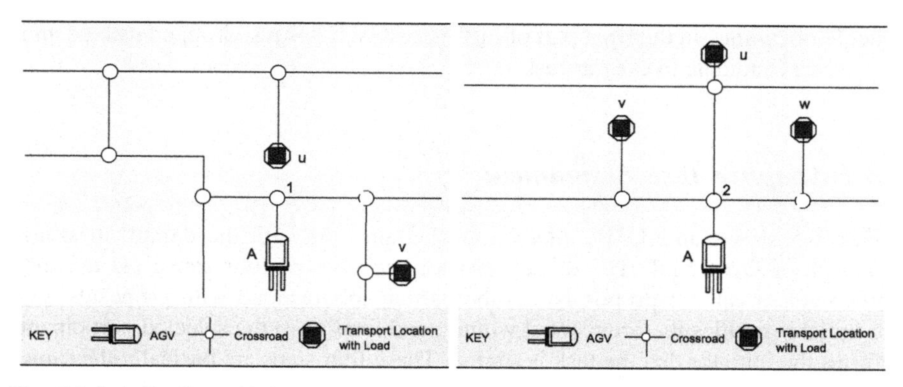

Fig. 6.4 *Left*: Dealing with local minima in FiTA. The attracting fields of transports u and v are proportional to the shortest path distance between AGV A and the transports. As such, AGV A will be guided toward transport v. *Right*: AGV A selects randomly between tasks v and w in node 2

A local minimum can also arise when the attracting fields and the repelling fields sensed by an AGV neutralize each other. Consider the situation on the left in Fig. 6.4. When AGV A computes its coordination field from node 2, the attracting fields of transport v and w may be equal and smaller than the field of transport u. In such a case, the AGV will select randomly one of the minimum fields to follow its route.

6.3 DynCNET Protocol

DynCNET is an extension of the well-known CNET protocol with "Dyn" referring to support for dynamic task assignment. DynCNET enables the agents to regularly reconsider the situation in the environment and adapt the assignment of tasks when circumstances change. The DynCNET protocol describes the behavior of AGV agents and transport agents to realize adaptive task assignment. This behavior is encapsulated by the agents' communication module.

DynCNET is an $m \times n$ protocol. An initiator that offers a task can interact with m participants, i.e., the candidate agents that can execute the task. On the other hand, each participant can interact with n initiators that offer tasks. As an example, consider the scenario shown in Fig. 6.6. In the AGV transportation system, an initiator corresponds with a transport agent that represents a task in the system and the participant corresponds with an AGV agent that can execute tasks. We denote the area where an initiator (or participant) searches for participants (or initiators) the *area of interest* of the initiator (or participant). The dotted circles in Fig. 6.6 show the current areas of interest of AGV A (top) and task x (bottom). For task x, there are currently two candidate AGVs to execute the task: F and G (AGV E is delivering a load). For AGV A on the other hand, there are three possible tasks to execute: u, v, and w. Because of the dynamics in the system, the set of candidate tasks (initiators) and agents that can execute a task (participants) can change over time. For example, in the right part of Fig. 6.7, AGV E has just dropped its load and becomes a candidate to execute task x.

6.3.1 Adaptive Task Assignment

Figure 6.5 shows an AUML interaction diagram [75] with the default message sequence of DynCNET. The default protocol consists of four steps: (1) the initiator sends a call for proposals; (2) the participants respond with proposals; (3) the initiator notifies the provisional winner; and finally (4) the selected participant informs the initiator that the task is started. These four steps are basically the same as in the standard CNET protocol. The flexibility of DynCNET is based on the provisional agreement between initiator and participant and the possible revision of the assignment of the task between the third and the fourth steps of the protocol.

To explain how agents can switch tasks when the conditions in the environment change, we use the UML state chart diagrams of Figs. 6.8 and 6.9. These state diagrams show the behavior of the participant and the initiator, respectively. When a task enters the system and it is ready to be executed (`task-ready`), the corresponding initiator enters the `Active` state in which it remains until the task is completed (`task–completed`) (see Fig. 6.9). As soon as a participant is `ready–to–work` it enters the `Working` state in which it remains until the task is executed (`ready`) (see Fig. 6.8). To explain the adaptability of DynCNET, we first look at the protocol from the perspective of the participant, then we look from the point of view of the initiator.

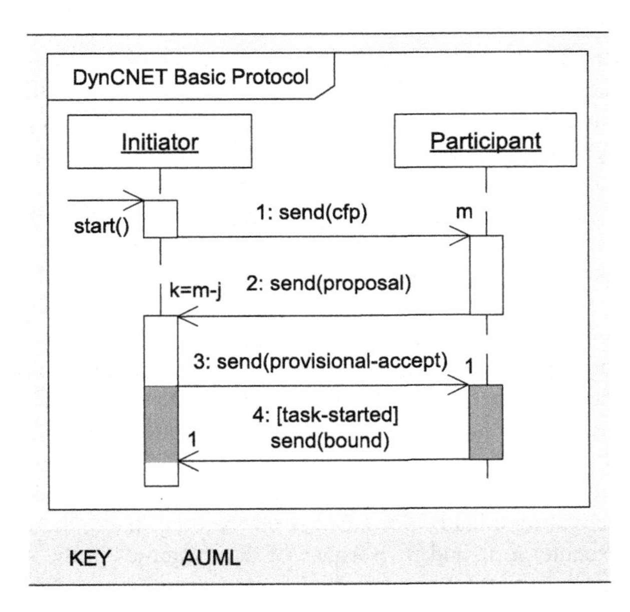

Fig. 6.5 High-level diagram of the DynCNET protocol. *Shaded zones* in the activation boxes represent periods in the protocol when agents can switch the provisional agreement

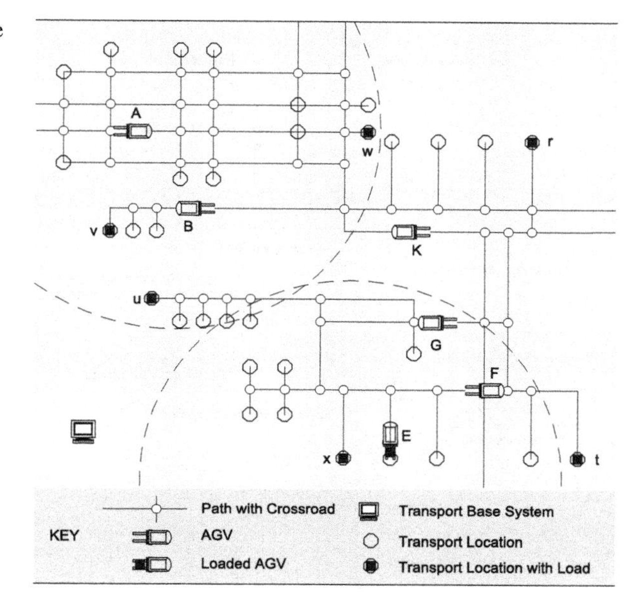

Fig. 6.6 Scenario to illustrate DynCNET. The *dotted circle* at the *top left* demarcates the current area of interest of AGV A. The *circle* at the *bottom* demarcates the current area of interest of task x

Switching Initiators. Consider the situation in Fig. 6.6 where we assume that AGV A has a provisional agreement to execute transport w. While AGV A drives toward the pick location of transport w, a new transport p enters the system, see the left part of Fig. 6.7. This new transport is an opportunity for AGV A to switch transport. DynCNET enables participants to switch initiators and exploit such opportunities. We use Fig. 6.8 for the explanation. When a participant is ready to execute a task, it enters the `Voting` state. As long as the participant has not received a provisional accept, it answers `cfp`'s with `proposals`. When the participant receives a `provisional-accept` message (step 3 in Fig. 6.5), it enters the

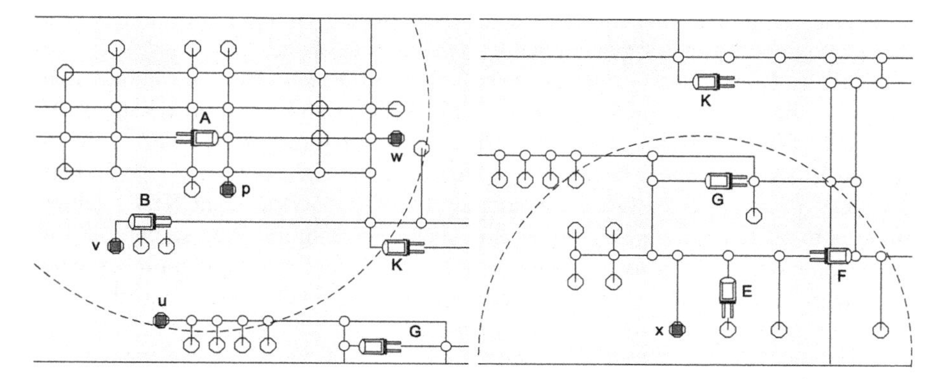

Fig. 6.7 *Left*: Task p provides an opportunity for AGV A to switch tasks. *Right*: AGV E becomes available for task x. The key is the same as in Fig. 6.1.

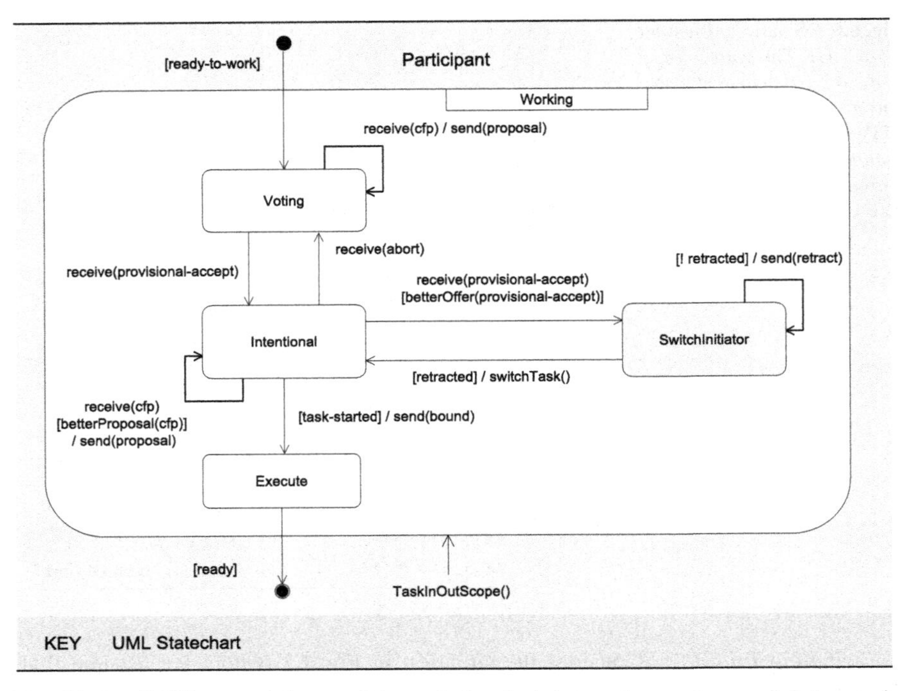

Fig. 6.8 DynCNET protocol for a participant. In the *shaded state*, the agent can switch the provisional agreement. The format of a state transition is *event [guard] / actions*

`Intentional` state. As soon as the participant starts the task (`task-started`), it sends a `bound` message to the initiator to inform the latter that the execution of the task is started. The participant is then committed to execute the task.[3] However, if a new opportunity occurs before the task is started, i.e., the participant receives a `provisional-accept` which is a better offer, the participant changes to the `Switch Initiator` state. The participant then retracts from the earlier provisional task assignment (`send(retract)`) and switches to the more suitable task (`SwitchTask()`) entering again the `Intentional` state.

Switching Participants. Consider the situation in Fig. 6.6 where we assume that transport x has a provisional agreement with AGV G and transport t with AGV F. While AGV G drives toward the pick location of transport x, AGV E drops the load it is carrying and becomes available, see the right part of Fig. 6.7. This new AGV is an opportunity for transport x to switch AGVs. DynCNET enables initiators to switch participants and exploit such opportunities. We use Fig. 6.8 for the explanation. As long as the initiator has not selected a participant to execute the task (`! haveWinner`), it sends periodically (`Timer()`) `cfp`'s to the par-

[3] The initiator's state changes from `Assigned` to `Executing` when it receives the bound message from the participant (see Fig. 6.9).

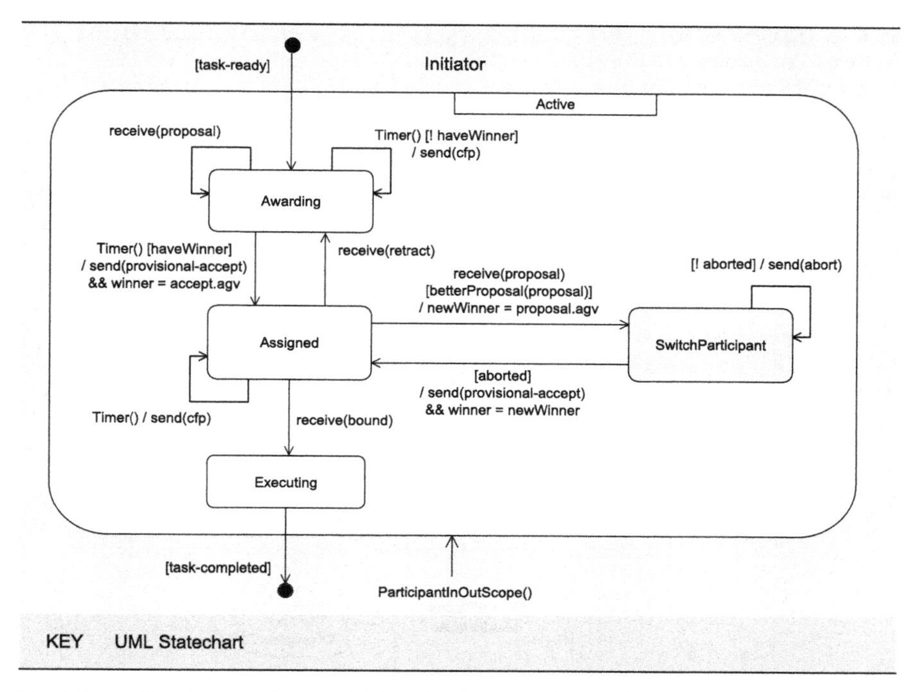

Fig. 6.9 DynCNET protocol for an initiator. In the *shaded state*, the agent can switch the provisional agreement. The format of a state transition is *event [guard] / actions*

ticipants in scope. Based on the received `proposals` from the participants, it selects a winner, sends a `provisional-accept` message (step 3 in Fig. 6.5), and enters the `Assigned` state. As soon as the initiator receives a `bound` message from the selected participant, it enters the state `Executing` in which the task is effectively started. However, if a new opportunity occurs before the task is started, i.e., the initiator receives a `proposal` from a participant which is better than the current provisionally accepted proposal, the initiator changes to the `Switch Participant` state. In this state the initiator sends an `abort` message to the provisionally assigned participant and subsequently sends a `provisional-accept` message to switch to the more suitable participant (`newWinner`).

6.3.2 *Monitoring the Area of Interest*

Participants use the function `TaskInOutScope()` to determine whether new tasks enter and leave their area of interest (see Fig. 6.8). Figure 6.10 shows a communicating components view of the elements involved in determining changes in the area of interest of an AGV agent.

To activate the `TaskInOutScope()` function, the communication component sends a monitoring request to the perception component. Perception registers for

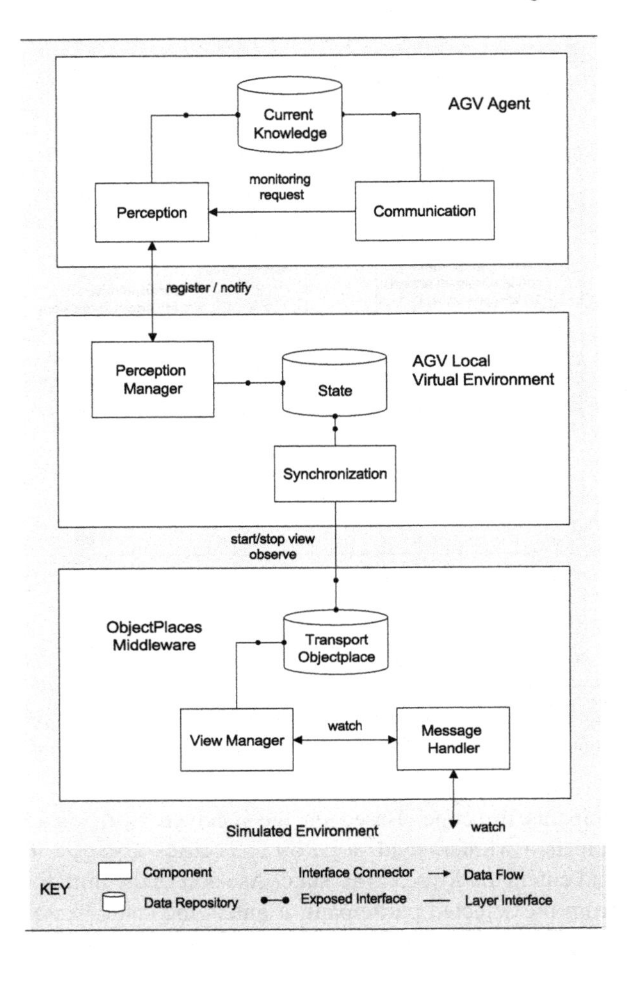

Fig. 6.10 The components involved in determining changes in the area of interest of an AGV agent

monitoring the AGV agent's area of interest with the perception manager. The perception manager requests the local state repository to monitor agents that enter and leave the specified area. The request is registered by the synchronization component which specifies a corresponding view and requests the view manager of ObjectPlaces to start building the view. The ObjectPlaces middleware constructs and maintains the view in the transport objectplace. Synchronization monitors the view objectplace and updates the state of the AGV local virtual environment when a change occurs. Changes are observed by the perception manager which notifies the perception component. Perception updates the agent's current knowledge which triggers the communication component when a task agent enters or leaves the agent's area of interest. As soon as the AGV agent picks a load, the perception component is informed and the monitoring process terminates. Similarly, the function `ParticipantInOutScope()` notifies the initiator when participants enter and leave its area of interest (see Fig. 6.9).

6.3.3 Convergence

A potential risk of DynCNET is that the assignment of tasks oscillates between participants and no tasks are executed. To ensure progress, both temporal and spatial windows are used in the protocol. Temporal windows are the time period used by the initiators between sending call for proposals in the awarding and the assigned state and the time period used by the participants between sending proposals in the voting and intentional state. Spatial windows are the size of the areas of interest for initiators and participants. We discuss temporal and spatial windows for a concrete AGV transportation system in Sect. 6.4. The tests show that the protocol converges for the selected time periods. However, additional research is required to formally prove convergence of the DynCNET protocol. A possible starting point to produce such a proof is described in [6]. In that paper, the authors formally prove the termination of an adapted CNET protocol.

6.3.4 Synchronization Issues

To avoid overloaded diagrams, we made abstraction of two synchronization problems in the description of the DynCNET protocol in Figs. 6.8 and 6.9. The first synchronization problem is related to a participant that has started executing a task, while an initiator has sent an abort message to that participant. The second synchronization problem is related to participants that leave the scope of interest of initiators. Appendix A.3 explains how these problems are solved.

6.4 Evaluation

To evaluate DynCNET and FiTA, we have applied both approaches in a simulated industrial AGV transportation system. After introducing the test setting, we present the main results of the tests, and we reflect on the test results.

6.4.1 Test Setting

All tests are performed on the map of an industrial AGV transportation system that is implemented by Egemin at EuroBaltic, a fishing processing center in Rugen, Germany, see Fig. 6.11. The size of the physical layout is 134×134 m. The map has 56 pick and 50 drop locations. We used a standard transport profile that Egemin uses for testing purposes. This profile generates 140 transports with a random pick location and a random drop location per hour real time. Each transport is assigned

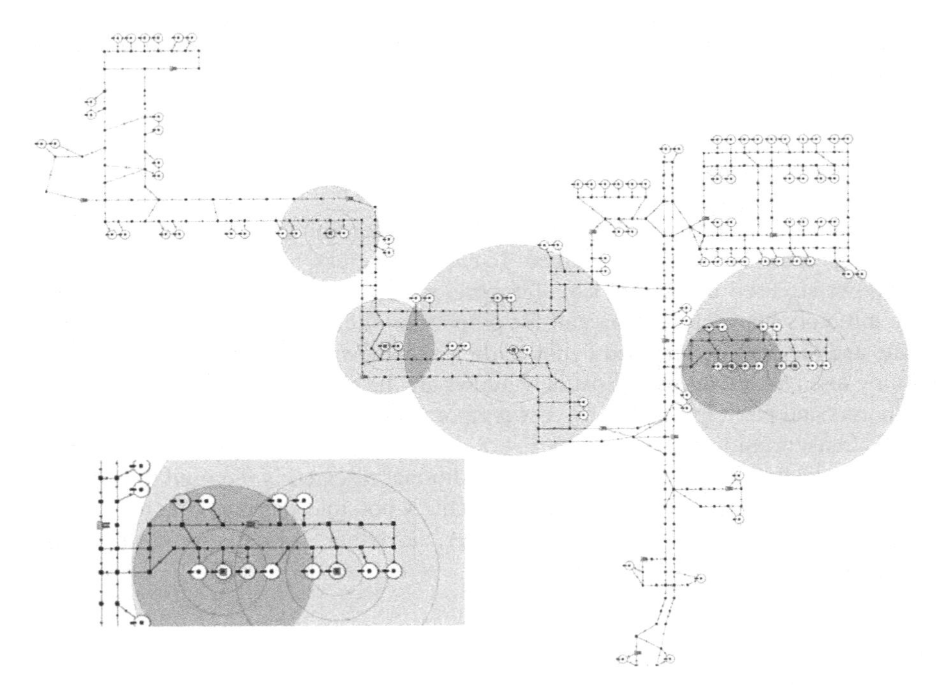

Fig. 6.11 Test map of the AGV transportation system that is implemented by Egemin at EuroBaltic. The snapshot is taken from a test with FiTA. The part at the *bottom left* zooms in on a small part of the map

a random priority that increases over time. In the simulation, we used 14 AGVs just as in the real setup. The average speed of driving AGVs is 0.7 m/s, while pick and drop actions take an average amount of time of 5 s. Every simulation was run for 200,000 timesteps, corresponding to approximately 4 h real time, i.e., one timestep represents 15 ms in real time. All displayed test results are average values over 30 simulation runs.

In the tests, we use standard CNET as a reference protocol. In CNET, an initiator calls for proposals and participants offer proposals to perform the task. When the initiator has received the proposals from all participants, it evaluates the proposals and assigns the task to the participant with the best offer. In the tests, a transport that enters the system is assigned as soon as possible to the most suitable AGV, i.e., an idle AGV for which the cost to reach the pick location is minimal. When transports cannot be assigned immediately, they enter a waiting status. All waiting transports are ordered by priority, and this priority determines the order in which transports are assigned. CNET is a static approach for task assignment which is comparable to schedule-based task assignment as traditionally used by Egemin.

Preceding to the tests, we determined the most suitable set of parameter values for the three task assignment approaches. Because of the constrained nature of the problem, in particular the restrictions imposed by the layout, for most parameters we could select a value within a range of possible values without significantly affecting the performance of the protocol.

6.4.2 Test Results

We focus on the evaluation of two important properties of the task assignment approaches: performance and robustness to message loss. Performance evaluation consists of two parts: communication load and completion of tasks over time. Communication load (number of messages sent per transport) is a crucial factor in multi-agent systems since decentralization of control requires more communication and thus additional bandwidth. Evaluation of the completion of tasks over time is important to demonstrate the flexibility of the task assignment approaches. To evaluate the completion of tasks over time, we measured reaction time (average waiting time per transport as a function of simulated timesteps) and throughput (number of finished transports as a function of simulated timesteps). Besides the test with a standard test profile, we have performed a stress test in which AGVs have to handle as quickly as possible a fixed number of transports from a limited number of locations. Robustness to message loss is another important criterion in decentralized systems, in particular in mobile systems that communicate via a wireless network. DynCNET is not robust to message loss since the protocol prescribes a particular sequence of message exchange. When a message gets lost, this sequence is disrupted and the interaction is blocked.[4] Given the multiple simultaneous interactions and the ongoing dynamics in the system, extending DynCNET with support for handling message loss is a non-trivial design task. Timeouts and confirmation messages are candidate tactics to develop such support, but they will introduce various design tradeoffs and have a significant impact on the protocol. Therefore, we have only tested robustness to message loss of FiTA. To evaluate the robustness to message loss, we have measured the reaction time and throughput for different degrees of message loss.

Since tasks are generated randomly and priorities are assigned randomly, we have verified the statistical significance for the main test results by calculating 95% confidence intervals. The confidence intervals are denoted with error bars in the figures. The relative small intervals indicate that the test results are sufficiently reliable.

6.4.2.1 Communication Load

To compare the communication load, we have measured the average number of messages sent per finished transport. Figure 6.12 shows the results of the test.

[4] In fact, some of the messages may get lost without blocking the interaction. For example, the protocol will not fail when a call for proposals message gets lost.

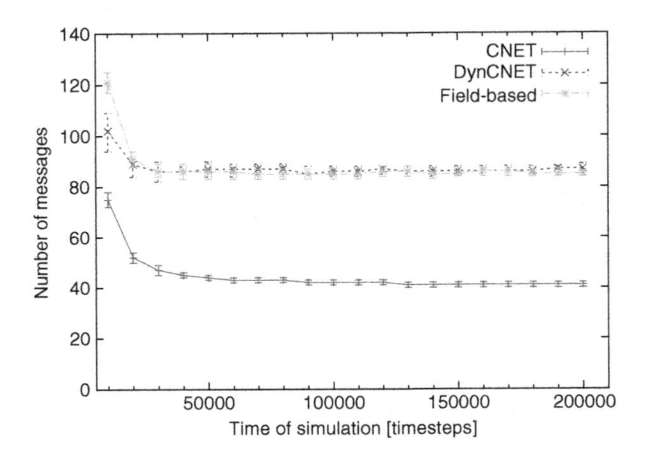

Fig. 6.12 Amount of messages being sent per finished transport

The number of messages of DynCNET and FiTA is approximately the same, while the communication load of CNET is about half of the load of the dynamic mechanisms. However, an important difference exists between the type of messages sent. Figure 6.13 summarizes the number of unicast and broadcast messages sent by the three mechanisms. For CNET, more than 90% of the communication is unicast messages. For DynCNET the balance unicast–broadcast messages is about 75–25%, yet for FiTA this balance is about 25–75%. This difference is an important factor for selecting appropriate communication infrastructure for a particular task assignment mechanism and vice versa.

6.4.2.2 Average Waiting Time

Figure 6.14 shows the test results for average waiting time for transports.

Average waiting time is expressed as the number of timesteps a transport has to wait before an AGV picks up the load. After a transition period of approximately 20,000 timesteps corresponding to approximately 20 min in real time, DynCNET and FiTA outperform CNET. The difference increases when time elapses. FiTA is

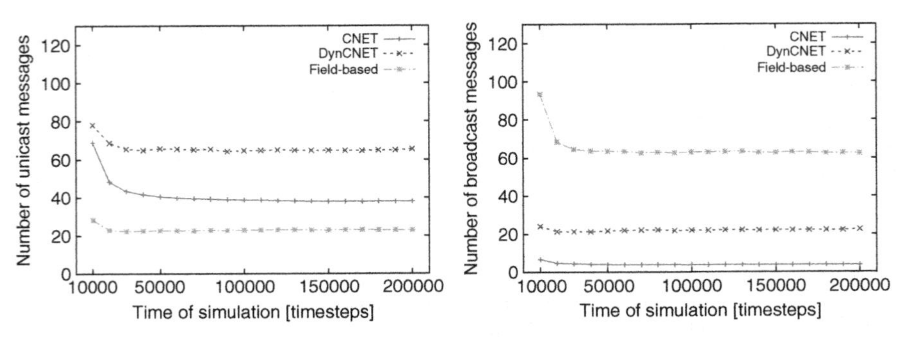

Fig. 6.13 *Left*: number of unicast messages. *Right*: number of broadcast messages

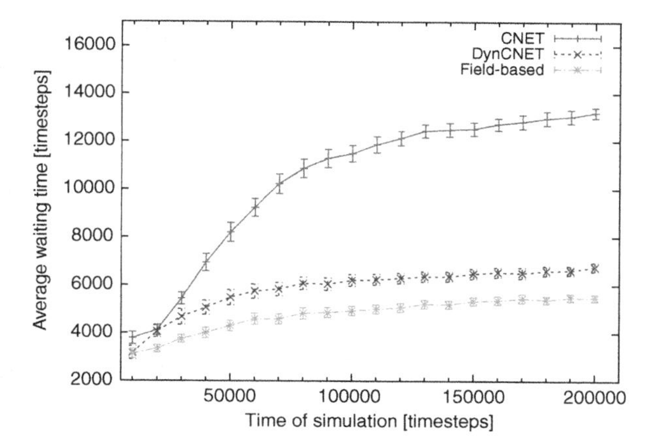

Fig. 6.14 Average waiting time per finished transport

slightly better than DynCNET over the full test range. A possible explanation is that idle AGVs in FiTA immediately start moving when they sense a field of a task, while in DynCNET AGVs only start moving after they are provisionally committed to execute a task.

In addition to the average waiting time for transports over time, we have measured the average waiting time per location. Figure 6.15 shows the results for the two dynamic task assignment approaches and CNET. CNET achieves a more equal distribution as the two adaptive task assignment approaches. In particular, the waiting times for pick locations 1–3 are significantly higher for FiTA and DynCNET. This drawback can be explained as follows: because these pick locations are far away from the main traffic in the warehouse, the chance an AGV will be close to the pick location is significantly lower. For FiTA, this decreases the chance for immediately attracting an idle AGV when a new transport pops up at the remote locations. Since the priority of a transport on the remote location gradually increases when the load is not picked, the field grows and eventually will attract an idle AGV. A

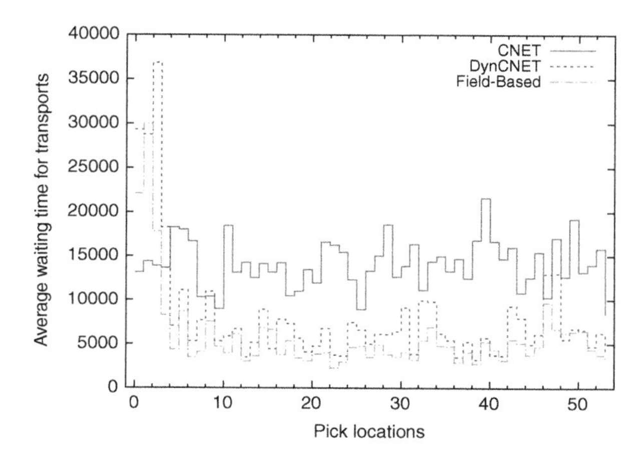

Fig. 6.15 Average waiting time per pick location

possible remedy to this problem is to increase the strength of fields of transports on isolated locations right from the moment the transport is created. For DynCNET, the situation is similar. The number of AGV agents in the area of interest of a transport agent located at a remote location will be lower and this will increase the waiting time for the transport. Increasing the area of interest of transport agents on isolated locations from the moment the transports enter the system is a possible solution to improve the distribution of waiting time over the transports.

6.4.2.3 Number of Finished Transports

Figure 6.16 shows the number of transports finished by each of the protocols during the test run. The results confirm the measures of the average waiting time per finished transport. DynCNET handles more transports than CNET, but less than FiTA. After 4 h in real time, on average, CNET has handled 380 transports, DynCNET 467 transports, and FiTA 515 transports. For the 467 executed transports of DynCNET, we measured an average of 414 switches of task assignments, 94.7% performed by transport agents and 5.3% by AGV agents.

Stress Test. In addition to the standard transport test profile, we have performed a stress test in which 45 transports are created at a limited number of locations in the beginning of the test. These transports have to be dropped at a particular set of destinations. The test simulates, for example, the arrival of a truck with loads that have to be distributed in a warehouse. The task of the AGVs is to bring the loads as quickly as possible to the right destinations. The transport test profiles for the three mechanisms were identical. Figure 6.17 shows the test results. The slopes of the curves of FiTA and DynCNET are similar but much steeper than the curve of CNET. The results demonstrate that CNET requires about 2.5 times more time to complete the 45 transports than the adaptive approaches.

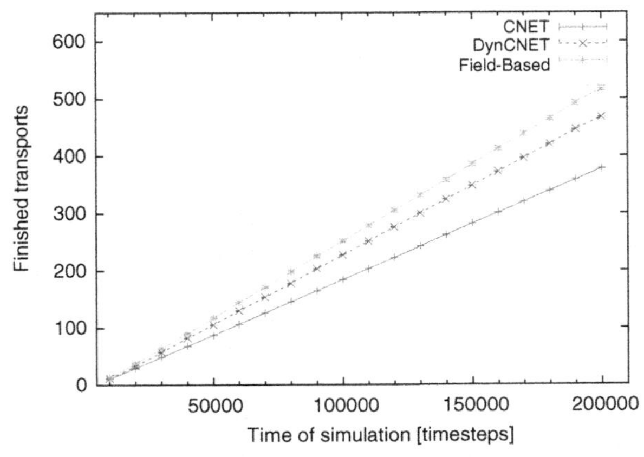

Fig. 6.16 Number of finished transports over time

s

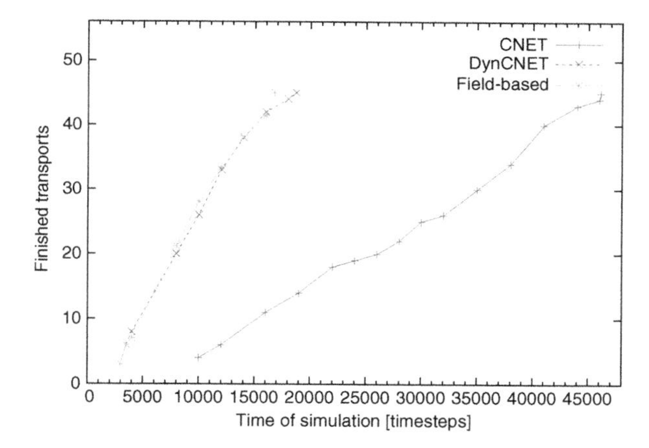

Fig. 6.17 Number of finished transports in the stress test

6.4.2.4 Robustness to Message Loss

As explained above, DynCNET is not robust to message loss and extending the protocol is a complex design task. Therefore, we have only tested robustness to message loss of FiTA. To demonstrate the robustness, we have measured the reaction time and throughput for different degrees of message loss. Figure 6.18 shows the average waiting time per finished transport for different percentages of message loss.

Figure 6.19 shows the corresponding number of finished transports over time. The test results show a graceful degradation of the performance of FiTA with increasing message loss. The average waiting time of transports systematically increases and the number of finished transports over time decreases with higher message loss rates. In practical AGV transportation systems, message loss is typically 1–2% with a maximum of 5%. The test results show that the impact of message loss of 2% is fairly limited. Even with 20% message loss, FiTA performs still better

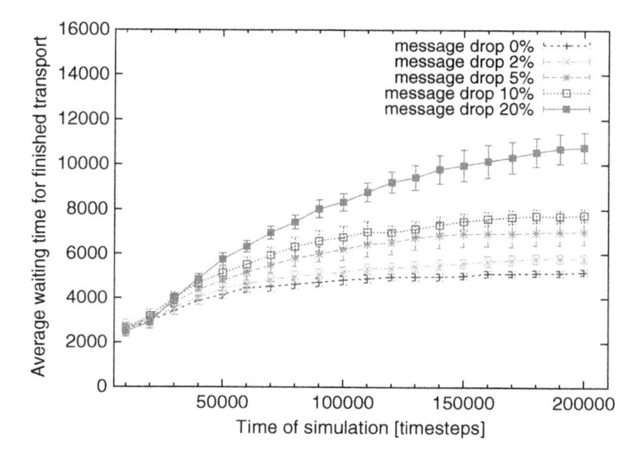

Fig. 6.18 Average waiting time for different degrees of message loss

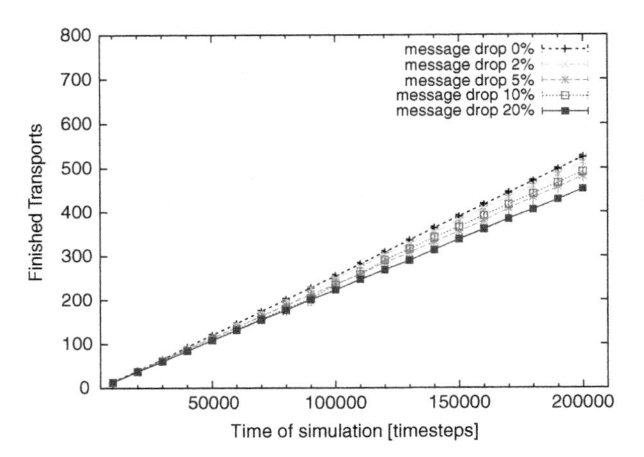

Fig. 6.19 Number of finished transports for different degrees of message loss

as CNET without message loss (compare the number of finished transports over time in Figs. 6.19 and 6.16).

6.4.3 Tradeoff Analysis

We now reflect on the test results and make a tradeoff analysis of the approaches for task assignment. First we zoom in on a number of important quality properties. Then we compare a number of engineering aspects.

6.4.3.1 Quality Attributes

DynCNET and FiTA have similar performance characteristics. Both outperform CNET on all performance measures; the cost is a doubling of required bandwidth. Since DynCNET explicitly defines the mechanism for agents to switch tasks, we expected that—when fine-tuned well—DynCNET would be able to outperform FiTA. However, the experiments show that this is not the case; at best DynCNET is able to equal the performance of FiTA. Figure 6.20 compares several additional quality attributes of the three task assignment approaches.

Flexibility. Flexibility refers to the agents' ability to adapt their behavior to dynamics that happen in the process of task assignment. Both DynCNET and FiTA support flexible assigning of tasks with delayed commencement, i.e., tasks that require a preceding effort before they can be executed. In FiTA, the choices of the participant agents are implicitly determined by the combination of the sensed fields. DynCNET provides explicit points of choice for initiators and participants. The points of choice are abstractly defined in the protocol and need to be instantiated according to the requirements of an application at hand. In the AGV application, agents use the priorities of tasks and the distance between AGVs and loads to switch

	Flexibility during delayed commencement	*Openness during delayed commencement*	*Robustness to message loss*
CNET	No: one shot assignment of tasks	Not supported	Requires additional support
FiTA	Yes: combination of fields determines the participants' choices	Inherent to the approach	Inherent to the approach
DynCNET	Yes: explicit points of choice for initiators and participants	Explicitly built-in	Requires additional support

Fig. 6.20 Summary of quality attributes of the three approaches

tasks. More advanced approaches can be considered, e.g., participants may (to some extent) favor tasks that are located near other tasks, increasing the chance to find a closely located task when the original assignment of tasks for some reason switches.

Openness. With openness, we refer to the agents' self-managing abilities to take into account other agents that enter and leave the system in the process of task assignment. Both DynCNET and FiTA support openness during delayed commencement, i.e., both mechanisms allow initiators to take into account new participants that become available and participants can participate in the assignment of new tasks that become available. Whereas FiTA inherently supports openness (the combination of fields adapts when fields disappear or new fields appear), the DynCNET protocol includes explicit functions (ParticipantInOutScope, InitiatorInOutScope) that notify initiators and participants when other agents enter or leave their current area of interest. Neither flexibility nor openness is supported by CNET.

Robustness. Robustness to message loss is the ability of a task assignment approach to withstand message loss (i.e., graceful degrade). In FiTA, the freshness of the perceived fields is taken into account to determine the attraction and repulsion of fields. When an agent misses an update of a field due to the loss of a message, the previous value of the field is used. Yet, to determine the combined field that guides the agent, less importance is given to older field values. As such, FiTA is (to some degree) robust to message loss. DynCNET (as CNET) on the other hand fails when a message gets lost and the prescribed sequence of messages is disrupted. As such, DynCNET requires additional support for robustness to message loss. Exception handling in protocol design is a non-trivial problem [105] and may significantly affect the properties of the protocol.

6.4.3.2 Engineering Aspects

Figure 6.21 compares a number of engineering aspects of the three task assignment approaches.

Engineering Mechanisms. No common engineering approaches are currently available for designing and developing FiTA. On the other hand, DynCNET allows to specify the behavior of the agents by means of common engineering diagrams such as interaction diagrams and state charts. We used UniMod [159] to design the DynCNET protocol as a state machine. UniMod enables to draw the state machine and export the diagram to an XML file. This XML file was used to interpret the state machine in the agent program.

Parameter Tuning. Parameter tuning is typically associated with stigmergy-based solutions such as FiTA. However, parameter tuning of DynCNET requires similar efforts as in FiTA. Examples are the range of interest of both types of agents, the growth rate to extend this range when no suitable candidates are found, the pace to send cfp and proposals. Our experiences indicate that a flexible agent-interaction protocol that deals with dynamics and change in the system also requires considerable efforts to tune parameters.

Type of Communication. A significant difference exists in the ratio of unicast–broadcast messages that are used in the three task assignment mechanisms. This difference is important for selecting appropriate communication infrastructure for a specific task assignment mechanism and vice versa. Note that an underlying network layer protocol such as TCP/IP is not a sufficient guarantee for robustness in DynCNET and FiTA since the interactions in both approaches involve broadcast communication.

	Engineering mechanisms	*Parameter tuning*	*Type of communication*
CNET	Use of common engineering diagrams	Limited number of parameters; tuning is easy.	Mainly unicast
FiTA	No common engineering approaches available	Various parameters; tuning requires considerable efforts	Mainly broadcast
DynCNET	Use of common engineering diagrams	Various parameters; tuning requires considerable efforts	Mainly unicast

Fig. 6.21 Summary of engineering aspects of the three approaches

6.5 Summary

In this chapter, we elaborated on the complex design problem of task assignment in multi-agent systems. We presented DynCNET and FiTA as two alternative approaches for adaptive task assignment in decentralized systems and applied them in an AGV transportation system. In FiTA, AGV agents follow fields in the AGV local virtual environment guiding them toward transports. DynCNET is an extension of the CNET protocol that allows agents to reconsider provisionally agreed assignments of transports when circumstances in the environment change. Both FiTA and DynCNET allow agents to adapt task assignment from the moment the transport enters the system until its execution is started. We have applied DynCNET and FiTA in a simulation industrial AGV transportation system that was implemented by Egemin. Our experiences yield the following conclusions:

- DynCNET and FiTA have similar performance characteristics and outperform CNET. Contrary to our expectations, DynCNET was not able to outperform FiTA.
- Whereas FiTA is inherently robust to message loss, DynCNET is not and requires substantial additional support to deal with message loss.
- Parameter tuning for DynCNET has similar complexity as for FiTA.
- DynCNET explicitly defines the task assignment process among the agents, while in FiTA task assignment is implicitly enclosed in the fields.

The tradeoff between support for robustness and engineering comfort—in particular the fact that DynCNET allows engineers to reason on the assignment of tasks—is an important criterion for selecting a task assignment approach in practice. We elaborate practical aspects of selecting an approach for task assignment in AGV transportation systems in Chap. 9.

Chapter 7
Evaluation of Multi-Agent System Architectures

Making the right architectural choices is crucial for successful development of a multi-agent system. Architectural evaluation allows examining a software architecture to determine whether it satisfies the important stakeholder requirements. Early evaluation of the software architecture enables adaptation of the architecture before the costs of correcting it become too high. The evaluation of a software architecture should involve an evaluation team and the stakeholders who have an interest in the architecture and the system that will be built from it.

In architecture-based design of multi-agent systems, we use the Architecture Tradeoff Analysis Method (ATAM) [46] for the evaluation of software architecture. ATAM is a structured method to examine whether a software architecture is suitable for the system for which it was designed. A suitable architecture is one that meets the stakeholder requirements, in particular the quality requirements. The main outputs of an architecture evaluation with ATAM are a prioritized set of quality attribute requirements, an analysis of architectural solutions to the main quality attributes, and a list of architectural tradeoffs and risks.

We start this chapter with a general introduction of the evaluation of a multi-agent system architecture with ATAM. Then, we explain in detail the ATAM evaluation for the case study. We conclude with a reflection on the experiences with using ATAM for the evaluation of a multi-agent system and a summary.

7.1 Evaluating Multi-Agent System Architectures with ATAM

Multi-agent systems are known for addressing quality attributes such as adaptability, robustness, openness, and scalability. However, for complex systems, stakeholders have various often conflicting requirements. For example, adaptability and performance may be major requirements for customers, configurability is important for deployment engineers, while reuse may be a prime concern of the project leader. The general goal of ATAM is to determine the tradeoffs and risks with respect to satisfying important quality attribute requirements. To evaluate a software architecture, ATAM focuses on important quality attribute scenarios identified by the

D. Weyns, *Architecture-Based Design of Multi-Agent Systems*,
DOI 10.1007/978-3-642-01064-4_7, © Springer-Verlag Berlin Heidelberg 2010

stakeholders. ATAM relies on both the architect and the architectural documentation (1) to identify architectural approaches and (2) to assess the way these approaches affect the quality attributes. The disciplined evaluation of the software architecture of a multi-agent system is invaluable to pinpoint the tradeoffs and risks implied by a multi-agent system architecture.

7.1.1 Architecture Evaluation in the Development Life Cycle

Figure 7.1 shows the part of the software development life cycle where architecture evaluation fits in.

Software architecture evaluation is typically done at an early stage of the design. Ideally, an ATAM starts when concrete descriptions of the quality attribute requirements and the software architecture are available, including a clear explanation of the main architectural design decisions. In practice, the descriptions are often vague and the ATAM serves as a means to refine and precise them.

One way to prepare an ATAM is by organizing a Quality Attribute Workshop (QAW) [19]. A QAW is a facilitated method that engages stakeholders to discover the driving quality attributes of a software-intensive system. During the QAW, a utility tree is constructed that helps to concretize and prioritize quality attributes. We give an example of a concrete utility tree for the case study in Sect. 7.2.3.

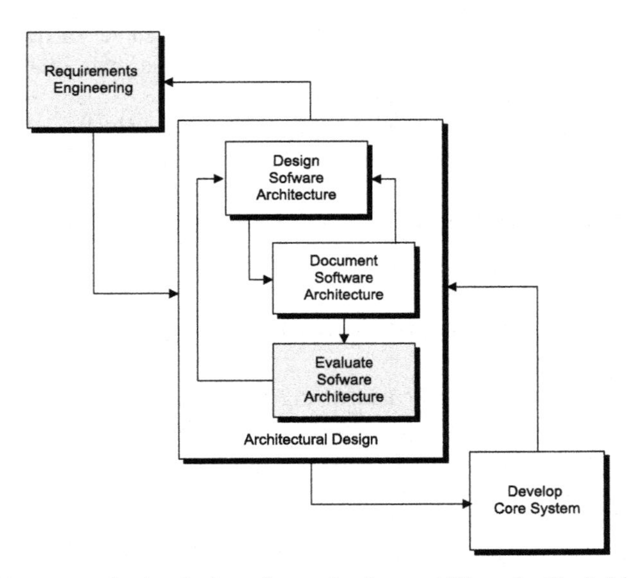

Fig. 7.1 Architecture evaluation in the software development life cycle. *Shaded boxes* represent the activities of interest in this chapter

7.1.2 Objectives of a Multi-Agent System Architecture Evaluation

Concrete objectives of a multi-agent system architecture evaluation with ATAM are

- To bring together the group of stakeholders interested in the multi-agent system architecture.
- To clarify the quality requirements. In particular, to clearly specify the architectural drivers that underlie the choice for a multi-agent system architecture, and the other important quality requirements of the system.
- To determine the relative importance of the quality requirements.
- To analyze how the multi-agent system architecture contributes to the realization of the main quality requirements.
- To detect possible risks in the software architecture. Risks are architecturally important decisions that have not been made or decisions that are not fully understood. An example of the former is the choice of a particular middleware for a multi-agent system architecture. An example of the latter is the lack of understanding of the communication overhead implied by a particular coordination mechanism.
- To analyze sensitivity points and tradeoff points in the architecture. A sensitivity point is an architectural decision that is critical for achieving a particular quality attribute. A tradeoff point is an architectural decision that affects more than one attribute; it is a sensitivity point for more than one attribute. An example of a sensitivity point is the choice for a particular type of agent organization to guarantee the required throughput of the system. An example of a tradeoff is when system performance improves with an increasing size of agent organizations, but robustness reduces.
- To evaluate the feasibility of building a concrete system based on the multi-agent system architecture.
- To finalize the architecture documentation.

It is important to notice that it is not the goal of an ATAM to precisely predict quality attribute behavior. Such prediction would require detailed information which is typically not available at an early stage of design. It is neither the goal to determine how problematic design decisions have to be tackled. The goal is to mitigate risks by bringing together the stakeholders to precisely determine the quality attribute requirements and analyze how the architectural decisions affect the achievement of these requirements.

7.1.3 Overview of the ATAM Activities

Figure 7.2 shows an overview of the activities of an ATAM.

The ATAM focuses on the identification of business drivers that determine the quality attribute requirements. The realization of the quality attribute requirements is based on the selection of architectural approaches, i.e., architectural tactics and

ATAM Workshop	
Introduction & ATAM Presentation Explanation of the importance of software architecture for software engineering and presentation of ATAM	**ATAM Leader**
Business Drivers Introduction of the project and presentation of the business drivers	**Project Manager**
Software Architecture Presentation of the software architecture. Focus on important view packets of module view, C & C view, and deployment view.	**Software Architects**
Architectural Approaches Presentation of important architectural approaches. The architects respond to questions of the evaluators in order to identify strenghts and weaknesses in the design.	**Software Architects**
Quality Attribute Tree Introduction of the quality attribute tree.	**Project Engineer**
Quality Attribute Tree The stakeholders discuss and revise the prioritized quality attribute scenarios.	**Discussion**
Architectural Approaches The architects explain how relevant architectural decisions contribute to the realization of the most important quality attribtute scenarios. Sensitivity points, tradeoffs, and risks are identified.	**Discussion**
Roundup and Workshop Closing	**ATAM Leader**

Fig. 7.2 Overview of the ATAM activities

patterns. During ATAM, the architectural decisions are analyzed to evaluate the strengths and weaknesses of the architecture.

7.2 Case Study

We explain the evaluation of a multi-agent system architecture for the AGV transportation system. In particular, we apply ATAM for one concrete application: an AGV transportation system for a tea processing warehouse.

We start this section with a brief introduction of the application, and we explain the business goals. Next, we give a general overview of the evaluation process. We zoom in on the QAW that was organized to elicit an utility tree. Then, we discuss the analysis of architectural approaches for two concrete quality attribute scenarios. We conclude with a brief discussion of follow-up activities of the ATAM.

7.2.1 AGV Transportation System for a Tea Processing Warehouse

In the tea warehouse application, bins with tea are stored in a warehouse and AGVs have to bring the full and empty bins to different tea-processing machines, such as machines for grinding, parching, and drying, and storage locations. The warehouse measures 75 × 55 m with a layout of approximately 6,000 nodes. The installation provides 12 AGVs that use navigation with magnet strips in the floor. There are 30 startup points for AGVs, i.e., points where AGVs can enter the system in a controlled way. AGVs use opportunity charging and a 11 Mbps wireless Ethernet is available for communication. Transports are generated by a warehouse management system. The average load is 140 transports/h, i.e., approximately 12 transports/AGV. Processing machines can be in two modes: low-capacity mode when machines ask for bins and high-capacity mode when bins are pushed to machines. Particular opportunities for optimization are double play (a double play is a combined transport consisting of a drop action in a predefined double play area by a specific vehicle and a pick action of a waiting load in the same area by the same vehicle), late decision for storage orders, and opportunity charging.

Important business goals for the tea warehouse transportation system are

- Flexibility with respect to storage capacity, throughput, and order profiles.
- Extendibility of the layout, production lines, and the number of vehicles.
- Reliability, i.e., 99.99% up-time, downtime may never cause production halt, and full tracing of quantities.
- Integration with ICT environment, wireless communication, security policy, and remote connectivity.

The installation is subject to a number of technical constraints, including backward compatibility with E'pia, the component framework developed by Egemin that provides basic middleware services for persistency, security, logging, etc., and compatibility with E'nsor, the low-level control software deployed on AGVs. Finally, the load of the wireless network is restricted to 60% of its full capacity.

7.2.2 Evaluation Process

The multi-agent system architecture was evaluated in a stage where the software architecture started to take shape. A prototype has been built and was tested. The main driver was to evaluate the software architecture before investing a major effort in the implementation.

Stakeholders Involved in the ATAM for the Tea Warehouse Application		
Stakeholder	**Role**	**Main Interests**
Software architect	Responsible for architectural design. Architectural decisions are tradeoffs among competing quality requirements.	Feedback on architectural design. Improve insight in the stakeholders' quality concerns.
Designer/Developer	Responsible for downstream design and implementation of the system.	Clarity and completeness of the design.
Project Manager	Responsibe project schedule and budget.	Clarity of the software architecture to allow allocation of resources to teams.
Service Engineer	Responsible for configuration and post-deployment activities.	Configurability and maintainability of the system.
Simulation Engineer	Responsible to set up a simulation environment to test the system.	Clarity of the design, and in particular correctness of the protocols.
Customer Representative	Represents a candidate purchaser of the system.	Costs of the system and usefulness of the system.
Project Engineer	Responsible for team formation and follow up of the project.	Clarity of the architecture structure to form teams, and manage milestones.

Fig. 7.3 Stakeholders involved in the ATAM of the AGV transportation system for the tea warehouse application, with their roles and main interests

The architecture evaluation was conducted by a team of three evaluators and nine stakeholders. Figure 7.3 shows an overview of the involved stakeholders with their main interests.

In preparation to the ATAM, three stakeholders together with one of the ATAM evaluators held a 4-day Quality Attribute Workshop. During the QAW a utility tree for the tea warehouse transportation system was developed. We discuss the QAW in the next section.

The ATAM itself took 1 day and followed the standard ATAM phases as described in Fig. 7.2. A key activity of the ATAM was the discussion on the mapping between the main quality attribute scenarios and the architectural approaches to achieve the quality attributes. This mapping illustrated the suitability of the architectural decisions as well as their weaknesses in the architecture and its documentation.

We discuss the mapping between two quality attribute scenarios and architectural approaches in Sect. 7.2.4.

7.2.3 Quality Attribute Workshop

To clarify system requirements before the software architecture was evaluated, a QAW was organized. The main activities of the workshop were

1. Identification of architectural drivers. The participants agreed upon the important quality attributes of the tea warehouse application and defined for each attribute a number of specific refinements in the context of the application. For example, for the quality attribute openness, two refinements were defined: "controlled adding and removing of an AGV" and "manual manipulation of an AGV."
2. Scenario brainstorming. Starting from the specification of the architectural drivers, the stakeholders generated concrete scenarios. Scenario generation is a key step in the QAW method. It is important to create well-formed scenarios that include a clear stimulus (i.e., the condition that affects the system), a response measure (i.e., the activity that results from the stimulus with a concrete measure by which the system's response will be evaluated), and a description of the environment in which the scenario takes place (i.e., the condition under which the stimulus occurred). An example of a scenario for openness is "If an operator removes or adds an AGV from the system in a controlled way, the rest of the system continues working." The role of the facilitator is crucial in this activity, she/he should require well-formed scenarios, ensure that scenarios are defined for all the refined architectural drivers that were identified, and propose to merge similar scenarios.
3. Scenario prioritization. Each scenario is assigned a ranking that expresses its priority relatively to the other scenarios. Scenarios are ranked according to the importance of the scenario to the success of the system and the difficulty to achieve the scenario.

The list of architectural drivers and the prioritized list of scenarios are key documents for the ATAM. The prioritized quality scenarios were structured in a utility tree. A utility tree groups the quality requirements in a tree structure in which high-level quality attributes are stepwise refined to concrete quality attribute scenarios. Concretely, a utility tree characterizes the driving attribute-specific requirements in a four-level tree structure where each level provides more specific information about important quality goals with leaves specifying measurable quality attribute scenarios.

Figure 7.4 shows an excerpt of the utility tree of the AGV transportation system. During the QAW, 11 different qualities and 34 concrete quality attribute scenarios were specified for the tea warehouse transportation system.

Each scenario is assigned a ranking that expresses its priority relative to the other scenarios. Prioritizing takes place in two dimensions. The first mark (high, medium,

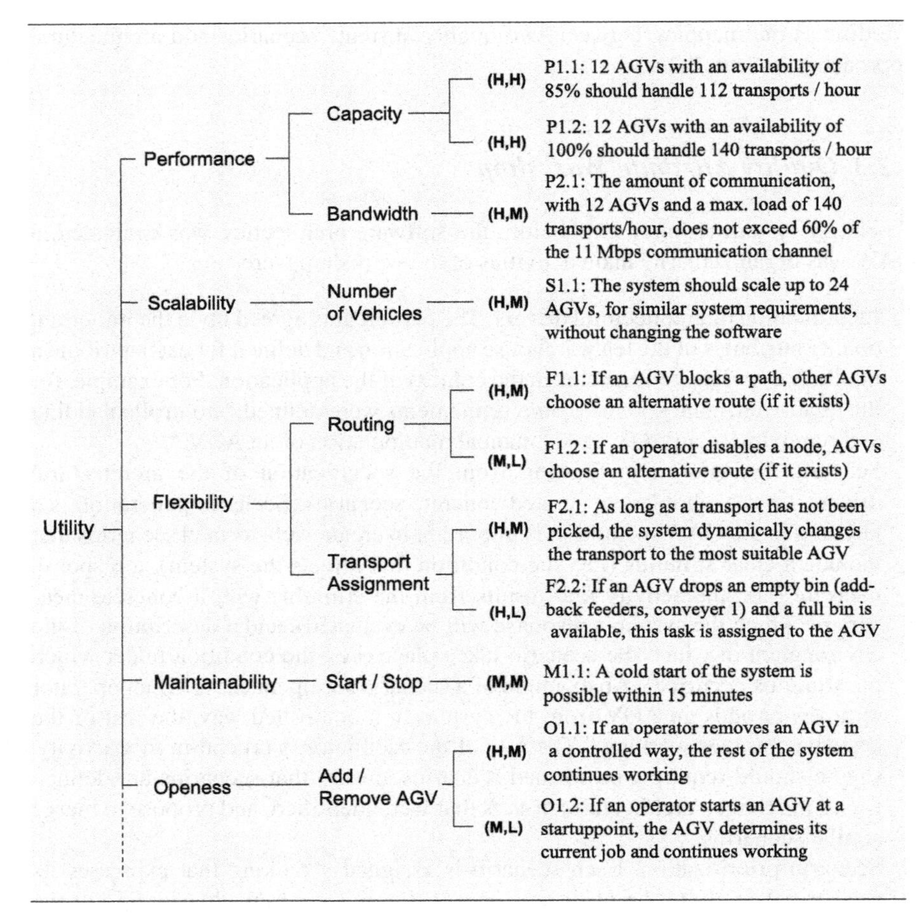

Fig. 7.4 Excerpt of the utility tree for the tea warehouse transportation system

or low) of each tuple refers to the importance of the scenario to the success of the system; the second to the difficulty to achieve the scenario.

At the ATAM workshop, minor changes were applied to the utility tree based on the discussion with the extended group of stakeholders.

7.2.4 Analysis of Architectural Approaches

During the ATAM workshop, the architectural approaches that address the high-priority quality attribute scenarios were identified and analyzed. A number of architectural risks (i.e., problematic architectural decisions), sensitivity points (i.e., architectural decision that involve architectural elements that are critical for achieving the quality attributes), and tradeoff points (i.e., architectural decisions that affect more than one attribute) of the software architecture were identified. The group of stakeholders discussed two particular quality attribute scenarios: one scenario concerning flexibility (transport assignment) and another scenario concerning performance (bandwidth usage). We give an overview of the results of the analysis of the two scenarios.

7.2.4.1 Architectural Analysis of Flexibility

Figure 7.5 shows an overview of the analysis of architectural decisions for the main quality attribute scenario of flexibility. The table shows the main architectural decisions (AD) that achieve the quality attribute scenario and specifies sensitivity points, tradeoffs, and risks associated with the architectural decisions. We give a brief explanation of the various architectural decisions:

AD 1. An agent is associated with each AGV and each transport in the system. To assign transports, multiple AGV agents negotiate with multiple transport agents. Agents continuously reconsider the changing situation, until a load is picked. The continuous reconsideration of transport assignments improves the flexibility of the system. However, it also implies a significant increase of communication. This was registered as tradeoff T1.

AD 2. For their decision making, agents take into account only local information in the environment. The most suitable range varies per type of information and may even vary over time for one particular type of information, e.g., candidate transports, vehicles to avoid collisions. The determination of this range for various functionalities is a sensitivity point. This sensitivity point was denoted as S1.

AD 3. The dynamic contract net protocol (DynCNET) for transport assignment is documented at a high level of abstraction. At the time of the ATAM, several important decisions were not taken yet. For example, how fast the agents need to reconsider the situation in the environment was an important choice that had not been made. Moreover, the difficulty of parameter tuning to ensure convergence and optimal behavior was unclear. This lack of clearness was registered as risk R1.

The architectural diagram show a high-level overview of the interactions between a transport agent and an AGV agent in the DynCNET protocol. The transport agent sends a *call-for-proposals* to *m* AGV agents and the AGV agent reacts with sending a *proposal* to *k* transport agents. Both agents select interaction partners based on the distance between the actual position of the AGV and the pick location of the load. When the transport agent has received all the proposals, it sends a *provisional-accept* message to the AGV agent with the best proposal. The agents then have a provisional agreement to execute the transport. While the AGV drives toward the load, the transport agent keeps sending call for proposals and the AGV agent reacts with proposals. Due to the mobility of the AGVs, the number of interaction partners may change (*k'* and *m'*, respectively). In case the AGV agent receives a *better offer* it *retracts*

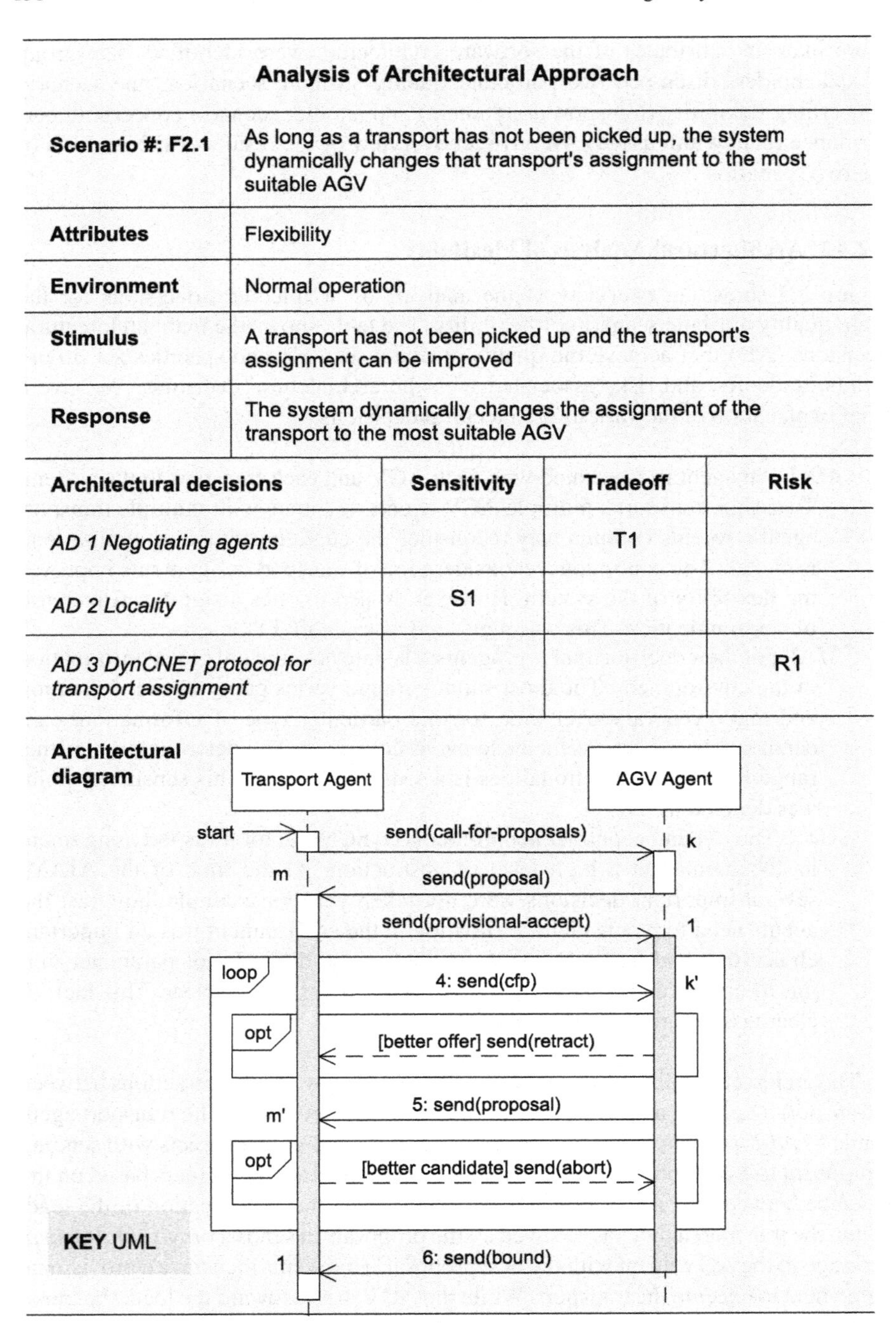

Analysis of Architectural Approach

Scenario #: F2.1	As long as a transport has not been picked up, the system dynamically changes that transport's assignment to the most suitable AGV		
Attributes	Flexibility		
Environment	Normal operation		
Stimulus	A transport has not been picked up and the transport's assignment can be improved		
Response	The system dynamically changes the assignment of the transport to the most suitable AGV		
Architectural decisions	**Sensitivity**	**Tradeoff**	**Risk**
AD 1 Negotiating agents		T1	
AD 2 Locality	S1		
AD 3 DynCNET protocol for transport assignment			R1
Architectural diagram			

Fig. 7.5 Analysis of architectural approaches with respect to flexibility

from the provisional agreement and switches transports. In case the transport agent finds a *better candidate* AGV, it *aborts* the provisional agreement and switches AGVs. The negotiation ends when the AGV with the provisional agreement picks the load and sends a *bound* message to the transport agent. We discuss task assignment with DynCNET in detail in Chap. 6.

7.2.4.2 Architectural Analysis of Bandwidth Usage

Figure 7.6 shows an overview of the analysis of architectural decisions for the main quality attribute scenario of bandwidth usage.

We give a brief explanation of the various architectural decisions:

AD 1. The AGV transportation system software is built on top of the .NET framework. This choice was not only a business constraint but also an evident choice since the E'pia library that is used for logging, persistence, security, etc., also uses .NET. The overhead induced by the choice for the point-to-point communication approach of .NET remoting was registered as a sensitivity point S2.

AD 2. Each AGV vehicle is controlled by an agent that is physically deployed on the machine. This decentralized approach induces a risk with respect to the required bandwidth for inter-agent communication. This was recorded as risk R2. An AGV agent can flexibly adapt its behavior to dynamics in the environment. AGVs controlled by autonomous agents can enter/leave the system without interrupting the rest of the system. However, flexibility and openness come with a communication cost. This tradeoff was noted as T2.

AD 3. The dynamic contract net protocol for transport assignment enables flexible assignment of transports among AGVs. Yet, the continuous reconsideration of transport assignment implies a communication cost. This tradeoff was denoted as T3.

AD 4. AGV agents use a two-phase deadlock prevention mechanism. AGV agents first apply static rules to avoid deadlock, e.g., agents lock unidirectional paths over their full length. These rules, however, do not exclude possible deadlock situations completely. If an agent detects a deadlock, it contacts the other involved agents to resolve the problem. Yet, the implications of the deadlock mechanism on the communication overhead are at the time of the ATAM not fully understood. This lack of insight was denoted as risk R3.

AD 5. The ObjectPlaces middleware uses unicast communication. However, some messages have to be transmitted to several agents, causing overhead. Support for multicast is possible, yet, this implies that the basic support of .NET remoting would no longer be usable. This potential problem was registered as sensitivity point S3 (see also S2).

The architectural diagram shows how multicast communication between agents is converted in point-to-point transmission. The scenario shows how the message of Agent 1 to Agent 2 and Agent 4 is converted in two separate messages.

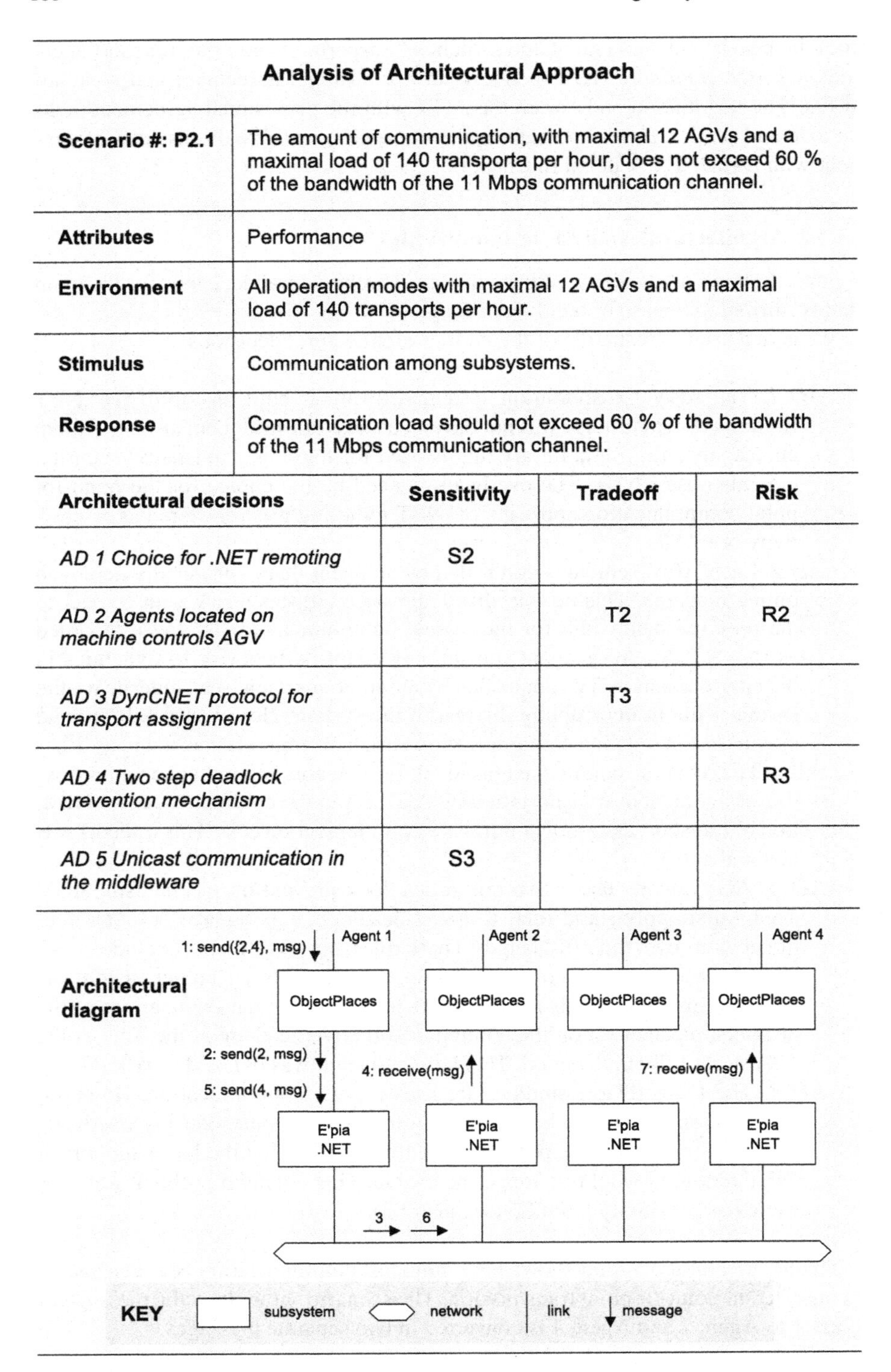

Analysis of Architectural Approach

Scenario #: P2.1	The amount of communication, with maximal 12 AGVs and a maximal load of 140 transporta per hour, does not exceed 60 % of the bandwidth of the 11 Mbps communication channel.		
Attributes	Performance		
Environment	All operation modes with maximal 12 AGVs and a maximal load of 140 transports per hour.		
Stimulus	Communication among subsystems.		
Response	Communication load should not exceed 60 % of the bandwidth of the 11 Mbps communication channel.		

Architectural decisions	**Sensitivity**	**Tradeoff**	**Risk**
AD 1 Choice for .NET remoting	S2		
AD 2 Agents located on machine controls AGV		T2	R2
AD 3 DynCNET protocol for transport assignment		T3	
AD 4 Two step deadlock prevention mechanism			R3
AD 5 Unicast communication in the middleware	S3		

Architectural diagram

Fig. 7.6 Analysis of architectural approaches with respect to bandwidth usage

Testing Communication Load. One important outcome of the ATAM evaluation was an improved insight into the tradeoff between flexibility and communication load. To further investigate this tradeoff, we conducted a number of tests after the ATAM workshop. Besides the simulation tests of the two approaches for transport assignment (see Chap. 6), we tested the efficiency of the middleware in the AGV application by measuring bandwidth usage of a system in a real factory layout.

Figure 7.7 shows the results of four consecutive test runs. We measured the amount of data sent on the network by each AGV and averaged this per minute to obtain the bandwidth usage relative to the bandwidth of a 11 Mbps IEEE 802.11 network. The first test (time 10–30 min) has three AGVs, of which two were artificially put in deadlock (a situation which is avoided in normal operation), because then the collision avoidance protocol is continually restarted and never succeeds. This is a peak load of the system. The second test (40–60 min) has three AGVs driving around freely. The third test (130–150 min) has five AGVs driving around freely. The fourth test (160–180 min) has five AGVs, all artificially put in deadlock. During the time in between test runs, AGVs were repositioned manually. On average, the bandwidth usage doubles when going from three to five AGVs. This is because the AGVs need to interact relatively more to avoid collisions. Based on these test results, Egemin experts consider the bandwidth usage acceptable for an extrapolation to 12 AGVs, taking into account that the maximal bandwidth usage should be less than 60% of the available 11 Mbps, and given that bandwidth optimizations were not applied yet.

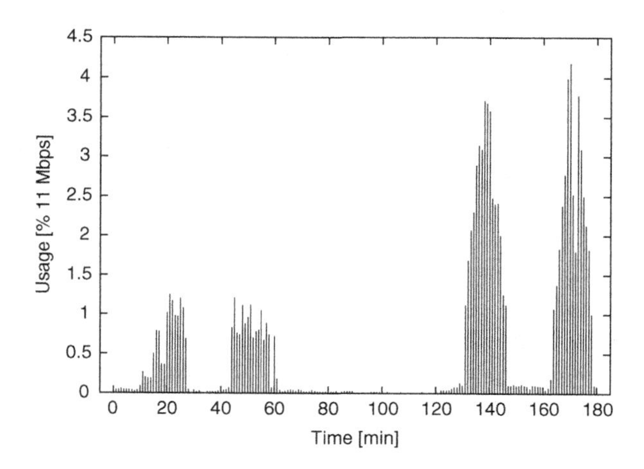

Fig. 7.7 Bandwidth usage in a test setting

7.3 Reflection on ATAM for Evaluating a Multi-Agent System Architecture

The ATAM workshop was a milestone in the design and development of the multi-agent system architecture for the AGV transportation system. For the first time, the assembled group of stakeholders discussed the software architecture in depth. Important results of the ATAM are

- Alignment of the business objectives with the architectural design. The business context provides the driving requirements for the system and the constraints within which the system has to be developed. On the other hand, a good understanding of the technical aspects allows the project decision makers to align their decisions with technical opportunities and constraints. Alignment of the business objectives with the architectural design is particularly important in the context of fielding a new architectural approach such as multi-agent system architecture.
- A list of prioritized quality attribute requirements, specified as scenarios which the stakeholders agreed on. Scenarios enforce stakeholders to precisely describe what the imported concerns of the system are. Bringing the stakeholders together and allowing them to express their concerns contribute to a better understanding of the required qualities and the tradeoff between different quality attributes. Prioritizing the scenarios helps to establish a collective vision on the relative importance of the different stakeholder concerns.
- Architectural evaluation is an incentive to provide a well-documented software architecture. A good architectural description that includes a rationale for the main architectural decisions is important for the evaluation of the architecture. However, afterward, the project will benefit from the architectural description as it serves as a key document for project organization and system development.
- Identification of risks, sensitivity points, and tradeoffs. The architecture evaluation provided insight into the strengths and weaknesses of the multi-agent system architecture. One interesting outcome of the ATAM was the clarification of the tradeoff between flexibility and communication load in the multi-agent system architecture. Although the architects were aware of this tradeoff, during the ATAM several architectural decisions were identified as risky and required further investigation. Field tests after the ATAM proved that the communication cost remains under control.
- A better insight into the impact on the software implied by the decentralization of control. Moving from a centralized architecture to a decentralized multi-agent system architecture implies a radical redesign which has a deep impact on the software.
- A better understanding of the importance of software architecture in the software engineering process. At the outset of the project, Egemin software architecture was not considered as an explicit part of the system's development process. The ATAM considerably contributed to improved architectural practice in Egemin.

A number of critical reflections about the ATAM were made as well:

- Performing a thorough and complete architectural evaluation during a 1-day ATAM is challenging. Clements et al. [46] suggest a three-day agenda for ATAM in which additional emphasis is given to scenario elicitation and prioritization and analysis of architectural approaches. Obviously, such an approach implies a proportional investment of the stakeholders.
- A QAW is an intense and demanding activity. Coming up with a comprehensive quality attribute tree is time consuming and at times tedious. The process can be

structured by letting each stakeholder prepare two or three scenarios in advance. These scenarios can serve as a good starting point and improve the efficiency for building up the utility tree. Involvement of an experienced facilitator who can guide scenario elicitation and building a utility tree is invaluable.

- The multi-agent system architecture for the AGV transportation system was developed with several automation projects in mind. However, the evaluation of the architecture was performed in the scope of a single automation project. Basically, ATAM is devised to evaluate a software architecture in a single project. The difference in scope sometimes complicated the evaluation since some architectural decisions were motivated by the product line nature of the software architecture.
- In preparation of the ATAM, the need for good tool support to document architectures became manifest. Without a proper tool, drawing architectural diagrams and building up the architectural documentation incur much overhead. Especially changing the documentation and keeping everything up-to-date (e.g., cross references and relations between different parts of the documentation) is tough and time consuming. Good tool support to document software architecture would be helpful.

7.4 ATAM Follow-Up and Demonstrator

The ATAM workshop initiated a number of follow-up activities. Several tests were conducted to investigate the main risks that were identified during the workshop. An extra analysis of risks and tradeoffs of the software architecture was performed with a reduced number of stakeholders. Finally, the architects finished the architectural documentation, and the evaluators presented the main workshop results. The report is available for download [30].

As a proof of concept, a demonstrator was developed for the decentralized AGV transportation system. The demonstrator with two AGVs is developed in .Net and supports the basic functionality for executing transport orders. The core of the action selection module of the AGVs is set up as a free-flow tree. A monitor enables remote access of the AGVs and generates a fusion view that represents the status of the local virtual environments of both AGVs. Figure 7.8 shows a snapshot of the AGVs in action with the fusion view.

Demonstration movies of the prototype implementation are available at the project Website [2].

7.5 Summary

Evaluation of the software architecture is an important activity in architectural-based design of multi-agent systems. It allows assessing the qualities of the multi-agent system without the need for a complete implementation. Architecture evaluation

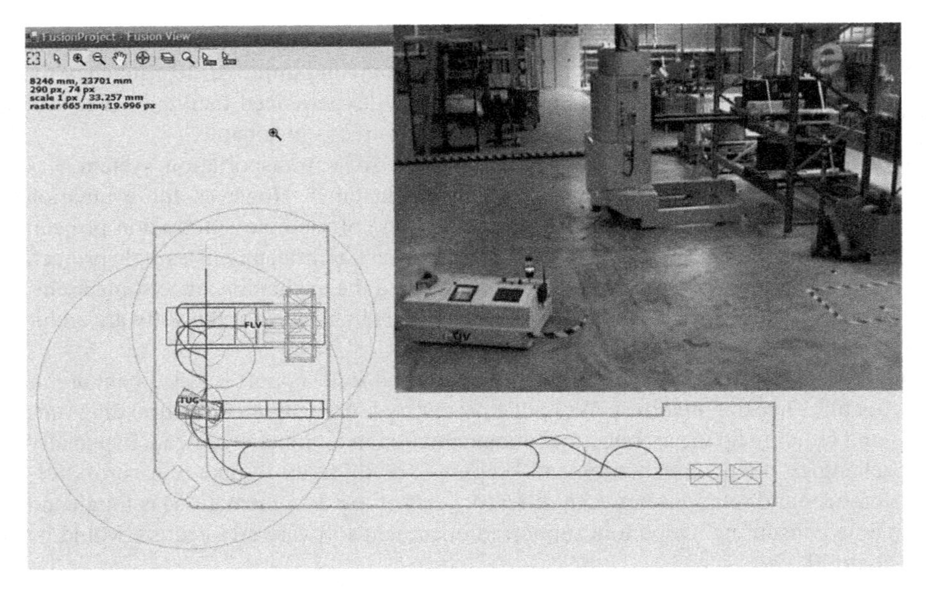

Fig. 7.8 Demonstrator with AGVs in action

brings the stakeholders together to discuss the software architecture of a system in which they have a common interest. It compels the stakeholders to define and prioritize the quality requirements of the system precisely. Architecture evaluation makes explicit causal connections between design decisions made by the architect and the qualities and properties in the software system. It allows determining the risks and tradeoffs with respect to satisfying important quality attributes of a multi-agent system, such as the tradeoff between adaptability and communication load. Architecture evaluation reveals the weak and strong points of the architecture, providing valuable feedback to the architects. This helps to avoid problems later in the development process when changes in the structure of the software are much harder to achieve or become too expensive.

In this chapter, we gave an overview of architecture evaluation of multi-agent systems and we showed the evaluation of a concrete multi-agent system architecture with ATAM. We explained the role of a QAW to identify the architectural drivers of the system and specify quality attributes that are ranked in a utility tree as concrete scenarios. We discussed the main activities of the ATAM workshop and zoomed in on the analysis of the architectural approaches for two important quality attribute scenarios. We pointed to a number of experiences with applying ATAM to the multi-agent system architecture. Finally, we discussed the main results of the architecture evaluation and referred to the implementation of a demonstration system that was developed.

Chapter 8
Related Approaches

Architecture-based design of multi-agent systems takes an architecture-centric perspective on the engineering of agent-based systems. The approach integrates multi-agent system concepts with state-of-the-art principles and methods of conventional software engineering. Although architectural design is considered as an explicit phase in several agent-oriented methodologies, none of them puts software architecture in the center of the engineering activities. This does not alter the fact that multi-agent system researchers have developed a body of knowledge on architectures for agent-based systems.

In this chapter, we discuss related approaches that explicitly consider the connection between software architecture and multi-agent systems. The discussion is divided into two main parts. We start by discussing related work on architectural approaches and multi-agent systems. Then, we explain related work on middleware support for distributed, decentralized applications. We focus on approaches that support the development of systems that are deployed in a mobile network. Additionally, we give a brief overview of related work on the case study that we used throughout this book: automated transportation systems.

It is important to notice that the overview is not intended to be complete; our goal is to give a representative overview of related research.

8.1 Architectural Approaches and Multi-Agent Systems

We start with discussing related work on architectural styles and multi-agent systems. Then, we explain related work on reference models and architectures for multi-agent systems.

8.1.1 Architectural Styles

In this section, we discuss related work on quality attributes and architectural styles for multi-agent systems.

Architectural Properties of Multi-agent Systems. In [149], Shehory presents an initial study on the role of multi-agent systems in software engineering, and in

D. Weyns, *Architecture-Based Design of Multi-Agent Systems*,
DOI 10.1007/978-3-642-01064-4_8,

particular their merit as a software architecture style. The author observes that the largest part of research in the design of multi-agent systems addresses the following question: Given a computational problem, can one build a multi-agent system to solve it? However, a more fundamental question is left unanswered: Given a computational problem, is a multi-agent system an appropriate solution? An answer to this question should precede the previous one, lest multi-agent systems may be developed where much simpler, more efficient solutions apply.

The author presents an initial set of architectural properties that can support designers to assess the suitability of a multi-agent system as a solution to a given problem. The properties provide a means to characterize multi-agent systems as a software architecture style. Properties include the agent internal architecture, the multi-agent system organization, the communication infrastructure, and other infrastructure services such as a location service, security, and support for mobility. Starting from this perspective, the author evaluates a number of multi-agent system frameworks and applications and compares them with respect to performance, flexibility, openness, and robustness.

Although the discussed properties are not unique to multi-agent systems, the author states that the combination of the properties results in systems that are suitable for solving a particular family of problem domains. Characteristics of these domains are distribution of information, location, and control; the environment is open and dynamically changing; and uncertainty is present. At the same time, the author points out that if only a few of these domain characteristics are present, it may be advisable to consider other architectures as solutions instead of a multi-agent system.

Almost a decade later, the majority of researchers in agent-oriented software engineering still pass over the analysis whether a multi-agent system is an appropriate solution for a given problem. We share the position of Shehory. In particular, (1) a designer should consider a multi-agent system as one of the possible architectural solutions to a problem at hand and (2) the choice should be driven by the characteristics of the problem domain and the quality goals of the system.

Organizational Perspective on Multi-agent Architectures. As part of the Tropos methodology [64], a set of architectural styles was proposed which adopt concepts from organization management theory [89, 43]. The styles are modeled using the $i^{\star}$ framework [179] which offers modeling concepts such as actor, goal, and actor dependency. Styles are evaluated with respect to various software quality attributes.

Proposed styles are joint venture, hierarchical contracting, bidding, and some others. As an example, the joint venture style models an agreement between a number of primary partner actors who benefit from sharing experience and knowledge. Each partner actor is autonomous and interacts directly with other partner actors to exchange services, data, and knowledge. However, the strategic operations of the joint venture are delegated to a joint management actor that coordinates tasks and manages the sharing of knowledge and resources. Secondary partner actors supply services or support tasks for the organization core.

Different kinds of dependencies exist between actors, such as goal dependencies, task dependencies, and resource dependencies. An example of a task dependency in a joint venture is a coordination task between the joint management actor and a principal partner. A particular kind of dependency is the so-called softgoal that is used to specify quality attributes. Softgoal dependencies are similar to goal dependencies, but their fulfillment cannot be defined precisely [43]. Kolp and colleagues [89] state that softgoals do not have a formal definition and are amenable to a more qualitative kind of analysis. Examples of softgoals in the joint venture style are "knowledge sharing" among principal partner actors and "added value" between the joint management actor and a principal partner actor. According to [43], a joint venture is particularly useful when adaptability, availability, and aggregability are important quality requirements. A joint venture is partially useful for systems that require predictability, security, cooperativity, and modularity. The style is less useful when competitivity is an important quality goal.

A softgoal dependency in Tropos has no clear definition; it lacks a criterion to verify whether the goal is satisfied or not. On the contrary, in architecture-based design of multi-agent systems, we use quality attribute scenarios to precisely specify quality goals. A scenario includes a criterion to measure whether the scenario is satisfied. In [88], Klein and colleagues describe "general scenarios" that allow a precise articulation of quality attributes independent of a particular domain. This allows to specify scenarios for architectural styles. Another difference relates to the way tradeoffs among quality requirements are handled. Whereas a utility tree allows prioritization of quality requirements to determine the drivers for architectural design, Tropos does not consider a systematic prioritization of quality goals such as a utility tree. In Tropos, a designer visualizes the design process and simultaneously attempts to satisfy the collection of softgoals for a system.

The assessment of architectural styles in Tropos is based on a set of quality attributes. Some of these attributes, such as availability and reliability, have been studied for years and have generally accepted definitions. Other attributes, such as cooperativity, competitivity, and aggregability, do not. Naming such attributes by themselves is not a sufficient basis on which to judge the suitability of an architectural style. The pattern language for situated multi-agent systems rigorously specifies the various architectural patterns and explains the rationale for the design of each pattern.

Architectural Evaluation of Agent-Based Systems. In [175], Woods and Barbacci study the evaluation of quality attributes of architectures of agent-based systems in the context of ATAM. The authors put forward an initial list of four relevant quality attributes for agent-based systems. The first attribute is performance predictability. Due to the autonomy of agents, it is difficult to predict the overall behavior of the system. The second attribute is security. Verifying authenticity for data access is an important concern of many agent-based systems. The third quality attribute is adaptability to changes in the environment. Agents are usually required to adapt to changes in their environment, including agents that leave the system and new agents that enter the system. The fourth attribute considered is availability. Availability of functionality is related to the presence of agents and other services in the system and their mutual dependencies.

To discuss quality attributes in agent-based systems, the authors propose a classification of agent-based systems. The classification abstracts from particular agent architectures, but focuses on the coordination among agents. The classification is inspired by previous work of Hayden and colleagues [69]. Example classes are matchmaker and broker that act as mediators between agents that provide services and agents that request for services. For each class, the authors define a set of quality attribute scenarios. Scenarios are formulated in a template form that consists of three parts: "may affect" describes the quality attributes that may be affected by the scenario; "implications" describes the risks or potential problems illuminated by the scenario; and "possible solutions" proposes ways to cope with possible risks. As an example, one of the scenarios of matchmaker is "provider fails after advertising service." This scenario may affect performance and reliability of the consumer of the service. A possible implication might be that the consumer blocks while it is waiting for the service, holding up the system. One possible solution is to let the consumer time out and notify the matchmaker.

The approach of Woods and Barbacci requires a decomposition of the agent-based system into primitive fragments that fit the generic-defined agent types (matchmaker, broker, etc.). Scenarios can then be specified based on the interaction between the identified fragments. However, the presented scenarios are generic and lack specificity. When applied to a real system such as the AGV application, scenarios should be further refined according to the domain-specific requirements and constraints. In addition, the scenarios only support the evaluation of communicative interactions between the agents. For some domains this may cover a significant part of the system. However, for other domains such as the AGV application, direct communication takes up only a small part of the system.

8.1.2 Reference Models and Architectures for Multi-Agent Systems

In this section, we discuss a number of representative reference models and architectures for multi-agent systems.

PROSA: Reference Architecture for Manufacturing Systems. In [178], Wyns and colleagues define a reference architecture as a set of coherent engineering and design principles used in a specific domain. PROSA (Product–Resource–Order–Staff Architecture) defines a reference architecture for a family of coordination and control applications, with manufacturing systems as the main domain. These systems are characterized by frequent changes and disturbances. PROSA aims to provide the required flexibility to cope with these dynamics.

The PROSA reference architecture [36, 160] is built around three types of basic agents: resource agent, product agent, and order agent. A resource agent contains a production resource of the manufacturing system and an information processing part that controls the resource. A product agent holds the know-how to make a product with sufficient quality; it contains up-to-date information on the product life cycle. Finally, an order agent represents a task in the manufacturing system; it is responsible for performing the assigned work correctly and on time. The agents exchange knowledge about the system, including process knowledge (i.e., how to perform a certain process on a certain resource), production knowledge (i.e., how to produce a certain product using certain resources), and process execution knowledge (i.e., information and methods regarding the progress of executing processes on resources). Staff agents are supplementary agents that can assist the basic agents in performing their work. Staff agents allow to incorporate centralized services (e.g., a planner or a scheduler). However, staff agents only give *advice* to basic agents; they do not introduce rigidity in the system.

The PROSA reference architecture uses object-oriented concepts to model the agents and their relationships. Aggregation is used to represent a cluster of agents that in turn can represent an agent at a higher level of abstraction. Specialization is used to differentiate between the different kinds of resource agents, order agents, and product agents specific for the manufacturing system at hand.

The specification of the PROSA reference architecture is descriptive. PROSA specifies the responsibilities of the various agent types in the system and their relationships, but abstracts from the internals of the agents. As a result, the reference architecture is easy to understand. The lack of a rigorous specification allows for different interpretations. An example is the use of object-oriented concepts to specify relationships between agents. Although intuitive, in essence it is unclear what the precise semantics is for notions such as "aggregation" and "specialization" for agents. What are the constraints imposed by such a hierarchy with respect to the behavior of agents as autonomous and adaptive entities?

An interesting extension of PROSA in which a virtual environment is exploited to obtain BDI (Believe, Desire, Intention [134]) functionality for the various PROSA agents is proposed in [73]. The approach avoids the complexity of BDI-based models and the accompanying computational load. In particular, the approach introduces the concept of "delegate multi-agent system." A delegate multi-agent system consists of light-weight agents which can be issued by the different PROSA agents. These ant-like agents explore a virtual environment that represents the underlying physical environment. The ants bring relevant information back to their responsible agent and put the intentions of the responsible agent as information in the virtual environment. This allows delegate multi-agent systems of different agents to coordinate by aligning or adapting the information in the virtual environment according to their own tasks. A similar idea was proposed by Bruecker in [35] and has recently further been elaborated by Parunak and Brueckner, see [124]. The use of the virtual environment in the work of [73] is closely connected to the virtual environment pattern of the pattern language for situated multi-agent systems. Verstraete and colleagues [161] propose a first step toward the integration of the BDI functionality provided by a delegate multi-agent system with the architecture of the cognitive agent that issues the delegate multi-agent system in the virtual environment.

Aspect-Oriented Agent Architecture. In [60], Garcia and colleagues observe that several agent concerns such as autonomy, learning, and mobility crosscut each other and the basic functionality of the agent. The authors state that existing approaches that apply well-known patterns to structure agent architectures—an example is the layered architecture of Kendall [84]—fail to cleanly separate the various concerns. This results in architectures that are difficult to understand, reuse, and maintain. To cope with the problem of crosscutting concerns, the authors propose an aspect-oriented approach to structure agent architectures.

The authors make a distinction between basic concerns of agent architectures and additional concerns that are optional. Basic concerns are features that are incorporated by all agent architectures and include knowledge, interaction, adaptation, and autonomy. Examples of additional concerns are mobility, learning, and collaboration. An aspect-oriented agent architecture consists of a "kernel" that encapsulates the core functionality of the agent (essentially the agent's internal state) and a set of aspects [86]. Each aspect modularizes a particular concern of the agent (basic and additional concerns). The architectural elements of the aspect-oriented agent architecture provide two types of interfaces: regular and crosscutting interfaces. A crosscutting interface specifies when and how an architectural aspect affects other architectural elements. The authors claim that the proposed approach provides a clean separation between the agent's basic functionality and the crosscutting agent properties. The resulting architecture is easier to understand and maintain and improves reuse.

The aspect-oriented agent architecture applies a different kind of modularization as described in the pattern language for situated multi-agent systems. Whereas a situated agent in the pattern language is decomposed in functional building blocks, Garcia and colleagues take another perspective on the decomposition of agents. The main motivation for the aspect-oriented agent architecture is to separate different concerns of agents aiming to improve understandability and maintenance. Yet it is unclear whether the interaction of the different concerns in the kernel (feature interaction [41]) will not lead to similar problems that the approach initially aimed to resolve. Still, crosscutting concerns in multi-agent systems are hardly explored and provide an interesting venue for future research.

Architectural Blueprint for Autonomic Computing. Autonomic computing is an initiative started by IBM in 2001. Its ultimate aim is to create self-managing computer systems to overcome their growing complexity [85]. IBM has developed an architectural blueprint for autonomic computing [1]. This architectural blueprint specifies the fundamental concepts and the architectural building blocks used to construct autonomic systems.

The blueprint architecture organizes an autonomic computing system into five layers. The lowest layer contains the system components that are managed by the autonomic system. System components can be any type of resource, a server, a database, a network, etc. The next layer incorporates manageability endpoints (touchpoints) that provide standard interfaces for managing the resources. Layer three constitutes of autonomic managers that provide the core functionality for self-management. An autonomic manager is an agent-like component that manages other software or hardware components using a control loop. The control loop of the autonomic manager includes functions to monitor, analyze, plan, and execute. Layer four contains autonomic managers that compose other autonomic managers. These compositions enable system-wide autonomic capabilities. The top layer provides a common system management interface that enables a system administrator to enter high-level policies to specify the autonomic behavior of the system. The layers can obtain and share knowledge via knowledge sources, such as a registry, a dictionary, and a database.

We now briefly discuss the architecture of an autonomic manager, the most elaborated part in the specification of the architectural blueprint. An autonomic manager automates some management function according to the behavior defined by a management interface. Self-managing capabilities are accomplished by taking an appropriate action based on one or more situations that the autonomic manager senses in the environment. Four architectural elements provide this control loop: (1) the monitor function provides the mechanisms that collect, aggregate, and filter data collected from a managed resource; (2) the analyze function provides the mechanisms that correlate and model observed situations; (3) the plan function provides the mechanisms that construct the actions needed to achieve the objectives of the manager; and (4) the execute function provides the mechanisms that control the execution of a plan with considerations for dynamic updates. These four parts work together to provide the management functions of the autonomic manager.

Although presented as architecture, in our opinion, the blueprint describes a reference model. The discussion mainly focuses on functionality and relationships between functional entities. The functionality for self-management must be completely provided by the autonomic managers. Obviously, this results in complex internal structures and causes high computational loads.

The virtual environment pattern in the pattern language for situated multi-agent systems provides an interesting opportunity to manage complexity. The virtual environment could enable the coordination among autonomic managers and provide supporting services.

A Reference Model for Multi-agent Systems. In [110], Modi and colleagues present a reference model for agent-based systems. The aim of the model is fourfold: (1) to establish a taxonomy of concepts and definitions needed to compare agent-based systems; (2) to identify functional elements that are common in agent-based systems; (3) to capture data flow dependencies among the functional elements; and (4) to specify assumptions and requirements regarding the dependencies among the elements.

The model is derived from the results of a thorough study of existing agent-based systems, including Cougaar [71], Jade [23], and Retsina [156]. The authors used reverse engineering techniques to perform an analysis of the software systems. Static analysis was used to study the source code of the software and dynamic analysis to inspect the system during execution. Key functions identified are directory services, messaging, mobility, inter-operability services, etc.

Starting from this data an initial reference model was derived for agent-based systems. The authors describe the reference model by means of a layered view and a functional view. The layered view is comprised of agents and their supporting framework and infrastructure which provide services and operating context to the agents. The model defines framework, platform, and host layers, which mediate between agents and the external environment. The functional view presents a set of functional concepts of agent-based systems. Example functionalities are administration (instantiate agents, allocate resources to agents, terminate agents), security (prevent execution of undesirable actions by entities from within or outside the agent system), conflict management (facilitate and enable the management of interdependencies between agents' activities), and messaging (enable information exchange between agents).

The reference model is an interesting effort toward maturing the domain. In particular, the reference model aims to be generic but does not make any recommendation about how to best engineer an agent-based system. Putting the focus on abstractions helps to resolve confusion in the domain and facilitates acquisition of agent technology in practice.

Yet since the authors have investigated only systems in which agents communicate through message exchange, the resulting reference model is biased toward this kind of agent systems. The concept of virtual environment as a means for information sharing and indirect coordination of agents is not supported. On the other hand, it is questionable whether developing one common reference model for the broad family of agent-based systems is desirable.

8.2 Middleware for Mobile Systems

In this section, we focus on middleware approaches that support the development of distributed, decentralized applications that are deployed in a mobile network. We focus on work related to views and coordination roles.

8.2.1 Work Related to Views

Tuplespaces Approaches. The first incarnation of a tuplespaces-based system is Linda [42]. Linda provides a shared collection of data tuples called the tuplespace, and a small set of tuple manipulation operations on the collection. Rowstron [142] augments the basic Linda model with asynchronous operations, as this scales better for distributed computing. A significant generalization of tuplespaces was the programmable tuplespace [116]. Programmable tuplespaces allow their behavior and operations to be programmed, by the specification of *reactions*. Basically, reactions are programs that are internally executed by the tuplespace. Reactions change the content of the tuplespace or the result of an operation in response to the execution of an operation. Examples of programmable tuplespaces are TuCSoN [116] and MARS [40]. Programmable tuplespaces allow the addition of any operation to a tuplespace, and in this way allow the encapsulation of the coordination rules in the tuplespace. Custom operations can be offered to the application on a higher abstraction level, and better tailored to the application's needs, than was previously possible.

Many-to-Many Invocation (M2MI) [81] is a middleware for distributed collaborative applications in mobile ad hoc networks that provide an object-oriented abstraction of broadcast messaging. Application components using the middleware can call a method on an "omnihandle" object, which calls the method on all the objects on connected nodes that implement the same interface as the omnihandle. The M2MI middleware relies on the unreliable broadcast mechanism in mobile ad hoc networks; it does not attempt to improve reliability, since any application in a mobile ad hoc network has to deal with joining and leaving nodes anyway.

M2MI's communication mechanism is less specific than that supported by views in ObjectPlaces. M2MI distinguishes receivers by the interface they implement and calls the method on objects deployed on all nodes within communication range. Views can distinguish nodes further based on the node constraint. Furthermore, objectplaces and views enforce uncoupling in time, while in M2MI the communicating components are tightly coupled in time: an object not in the network at the time an omnihandle call is executed is not able to receive it.

EgoSpaces [78] is an extension of LIME [112]. EgoSpaces and LIME are coordination middlewares for mobile ad hoc networks based on a tuplespaces approach, but augmented with support for mobility. Both in LIME and in EgoSpaces, each application component has a personal tuplespace that moves as the component moves. In LIME, operations to put, take, and read tuples from a component's tuplespace are automatically executed on tuplespaces of application components on the same and on connected nodes. So, all application components on connected nodes form an opportunistically shared tuplespace and use it to coordinate.

EgoSpaces offers similar functionality as LIME, but application components can select in a fine-grained way on which tuplespaces and on which nodes the tuplespace operations are executed, by defining an abstraction that is also called a view. A view is set up by EgoSpaces using a declarative specification of which nodes and tuplespaces are to be included when executing the operation. This specification is roughly similar in expressivity as the specification for views defined in ObjectPlaces. Besides reading and taking tuples from tuplespaces in the view, application components can also register so-called reactions on a view, which allow an application component to be notified, e.g., when a tuple enters or leaves the view.

The main differences between EgoSpaces' views and ObjectPlaces' views are

1. Views in ObjectPlaces are purely observational, i.e., a view is a local collection of copies of objects contained in objectplaces selected by the view specification. At the cost of some overhead, EgoSpaces also allows application components to remove tuples from tuplespaces in an EgoSpaces' view.
2. In EgoSpaces, a view always has the interface of a tuplespace, i.e., it supports operations to remove and read tuples by template matching. In ObjectPlaces, the representation of the view can be tailored to an application component's wishes: it can be a sorted collection, an accumulation to a value, or something else. As such ObjectPlaces' views can be seen as a realization of the idea of context-sensitive data structures presented in [126]. Context-sensitive data structures are basically abstract data structures, but whose content is defined by data available on connected nodes in the network.

The second point is a clear advantage for the application developer, as the views that he or she can work with can be tailored much better to how they are used, instead of being fixed to a rigid tuplespace-like interface. Also, view representations can be reused across applications, which potentially allows the construction of a library of reusable view representations.

As for the first point, in ObjectPlaces, views are kept purely observational entities, supporting coordination by information exchange only. In our experience, more complex coordination, which necessitates the manipulation of objectplaces on other nodes, requires full-fledged and application-specific protocols to be executed between participating nodes. Coordination roles support such protocol-based coordination.

TOTA (Tuples On The Air) [106] is a middleware that provides applications with the abstraction of a self-maintaining distributed tuple. Each node in the network hosts a tuplespace. A distributed tuple is propagated to nearby nodes and can be changed with each propagation according to an application-specific rule (e.g., counting the number of hops from the root). This tuple is then maintained by the middleware as the network changes.

TOTA is mainly used to support coordination by so-called gradient fields. A distributed tuple can be seen as a virtual field (compare to a magnetic or electric field) that has either an attractive or a repulsive influence on other entities. By a careful choice of the gradient fields that each entity emits, and its influence on other entities, gradient fields can be used for coordination that involves the coordinated motion of various kinds of objects in a metric space. In a software system, gradient fields are virtual and the strength, origin, and other properties of emitted fields must be transmitted over the network; TOTA's distributed tuples offer direct support for transmitting and maintaining gradient fields in mobile ad hoc networks.

A distributed tuple in TOTA can be seen as the inverse of a view in ObjectPlaces: instead of gathering objects from neighboring nodes, a distributed tuple spreads objects (or tuples) to neighboring nodes. An important difference is that a view is specific for every observer, while a distributed tuple is the same for all observers. In other words, while in TOTA the "sender" of a message (the component that adds a distributed tuple) determines both who it reaches and what the content is, using views the "receiver" of a message (the component that builds a view) can determine both content and representation. Views and distributed tuples thus represent two sides of information exchange; distributed tuples are better suited for information dissemination, while views are better suited for information gathering. In terms of support for the application developer, both approaches are complementary.

8.2.2 Work Related to Coordination Roles

Source-Initiated Context Construction in Mobile Ad Hoc Networks (SICC). In [79], the authors integrate the EgoSpaces middleware [78] with support of context awareness. The authors propose communication constructs and a protocol necessary to support context-aware interactions among mobile nodes. In SICC, a reference host (i.e., the host building the context) specifies the context over which it would like to operate but does not need to know the identities of the hosts in the context. Context information is defined as an abstraction of network properties. Specifying a context includes the definition of constraints that include a distance definition and a maximum allowable distance that may be based on a simple hop counter or may take into account the dynamic properties of the network such as latency. The middleware guarantees the application that a message sent to its context is received only by hosts belonging to the context and that all hosts belonging to the context receive the message. Furthermore, the infrastructure maintains the context based on the context definition.

SICC is mainly concerned with setting up and maintaining a group of nodes that comply to a context definition (i.e., a constraint) similar to group formation and views in ObjectPlaces. The middleware supports send–reply interactions among the reference node and the nodes in the context. However, protocol-based interaction with coordination roles as first-class entities as provided in ObjectPlaces is not supported in SICC.

Coordination Language Facility (CLF). CLF [10] offers support for the development of distributed object-oriented applications. It considers objects as resource managers and defines a set of performatives, i.e., primitive actions to manipulate an object's resources. Coordination between objects is expressed by scripts, which manipulate the resources by invoking the performatives (such as Inquiry, Confirm) on the different objects. Scripts are executed by coordinators that are responsible for guaranteeing the overall coherent behavior of the objects they coordinate. A standard two-phase commit protocol is used to guarantee consistency among the objects a coordinator controls.

CLF is not targeted toward mobile applications. However, CLF is mentioned because it allows a kind of protocol-based coordination, by allowing coordinators to execute long-lived negotiations with coordinated objects using the performatives. CLF does protocol-based coordination using generic protocol steps (i.e., two-phase commit) and a standard set of messages to manipulate resources (i.e., the performatives). Objectplaces are similar to the resources in CLF, since an objectplace also provides a generic interface to a number of resources. However, ObjectPlaces does not provide a standard set of protocol steps that allow the developer to deal with mobility easier.

CoorSet. [82] presents the coordination model CoorSet, not specifically targeted toward mobile applications, that is based on so-called associative broadcast. Components using associative broadcast include a "target set specification" with each message that they send. The target set of receivers is determined for each message as the set of receivers whose local state satisfies the target set specification. The sender does not know the membership of this set. The state of a component is specified as a so-called profile, which specifies the visible current state of the component. Profiles are implemented as sets of attribute–value pairs. The target set for a message is determined by a predicate called a selector, which is evaluated against the profile of each component. The message is received only by those components for whose profiles the selector evaluates to true. This allows targeting of messages to subsets of components that have a desired state, without the sender knowing the membership of that set.

A node constraint in ObjectPlaces is similar to CoorSet's selector, and the node properties repository is similar to CoorSet's profiles. CoorSet allows a component to reach another component based on the component's properties, while in ObjectPlaces a component can reach another component by a constraint on node properties and the name of a coordination role.

The main difference is that in CoorSet, each message is sent with its own selector, so no relation is maintained explicitly between the different messages in a protocol. If interaction partners change, this is not monitored and signaled by CoorSet, but must be handled by the application developer. In short, CoorSet does not set up and maintain interaction sessions, but concentrates on the uncoupled sending and receiving of individual messages; as such, it has many similarities with publish/subscribe systems.

Group Communication Systems. Groups of interacting coordination roles in ObjectPlaces can be compared to groups in group communication systems (GCS). Originally, GCS have been studied in fixed distributed systems and later in mobile ad hoc networks [139, 98]. The goal of GCS is to set up a group of application components in a distributed system, such that each member has a consistent (i.e., identical) view on the group's members and such that each member can communicate with all group members.

A GCS group is set up based on a commonly known group identifier. Application components join the group by sending a join message to the group identifier and can leave from the group in the same way. In contrast, a group of interacting coordination roles in ObjectPlaces are set up based on a node constraint, and coordination roles or application components do not have to join a group explicitly to take part in an interaction session. The latter is a more declarative and flexible way of setting up a group of interacting components. It allows the definition of a group in terms of any property of the members, instead of only on the basis of identity.

In ObjectPlaces, a group of coordination roles in an interaction session is a more fine-grained concept than a group in GCS: the latter is long-lived, i.e., each member is likely to initiate many interaction sessions during the lifetime of the group, while groups of interacting coordination roles are short-lived, i.e., for the duration of one interaction session. In other words, in one GCS group many interaction sessions are in progress concurrently.

As a result, GCS groups do not separate interaction sessions. ObjectPlaces instantiates a new coordination role for each interaction session in which an application component takes part. This enables the middleware to encapsulate tedious tasks with regard to session management, such as routing incoming messages to the appropriate coordination role for each interaction session. So, the application developer can for the most part make abstraction of the fact that different interaction sessions are executing concurrently.

Another difference is that in a group of interacting coordination roles, the role that has a complete view on the group's members is the initiator, whereas in GCS groups all members have a complete view of the group. For a group of coordination roles, this is reasonable since an interaction session is always started by one component concerning one particular "subject" that concerns the initiating component (e.g., the assignment of a particular task).

8.3 Scheduling and Routing of AGV Transportation Systems

The control of AGVs is subject of active research since the mid-1980s. Most of the research has been conducted in the domain of AI and robotics. Recently, a number of researchers have applied multi-agent systems, yet most of this work is applied in small-scale projects.

8.3.1 AI and Robotics Approaches

The problems of routing and scheduling of AGVs are different from conventional path finding and scheduling problems. Scheduling and routing of AGVs are a time-critical problem, while a graph problem usually is not. Besides, the physical dimensions of the AGVs and the layout of the map must be taken into account.

Roughly spoken, three kinds of methods are applied to solve the routing and scheduling problem. Static methods use a shortest path algorithm to calculate routes for AGVs, see, e.g., [48]. In case there exists an overlap between paths of AGVs, only one AGV is allowed to proceed. The other AGVs have to wait until the first AGV has reached its destination. Such algorithms are simple, but not efficient. Time-window-based methods maintain for each node in the layout a list of time windows reserved by scheduled AGVs. An algorithm routes vehicles through the layout taking into account the reservation times of nodes, see, e.g., [87]. Dynamic methods apply incremental routing. An example algorithm is given in [157]. This algorithm selects the next node for the AGV to visit (toward its destination) based on the status of the neighboring nodes (reserved or not) and the shortest travel time. This is repeated until the vehicle reaches its destination. Measurements show that the algorithm is significantly faster than non-dynamic algorithms, yet the calculated routes are less efficient.

A number of researchers have investigated learning techniques to improve scheduling and routing of AGVs, see, e.g., [130, 103]. This latter work applies reinforcement learning techniques and demonstrates that the approach outperforms simple heuristics such as first-come-first-served and nearest-station-first.

Contrary to the decentralized approach we have described in this book, traditional scheduling and routing algorithms usually run on a central traffic control system from where commands are dispatched to the vehicles [133]. Moreover, most approaches are intended to find an optimal schedule for a particular setting. Such approaches are very efficient when the tasks are known in advance as, for example, the loading and unloading of a ship in a container terminal. In our work, scheduling and routing are going concerns, with AGVs operating in a highly dynamic environment.

8.3.2 *Multi-Agent System Approaches*

Pallottino and colleagues [120] present a decentralized approach for collision-free movements of vehicles. In this approach, agents use cognitive planning to steer the AGVs through the warehouse layout. Berman and colleagues [24] discuss a behavior-based approach for decentralized control of automatic guided vehicles. In this work, conflict resolution with respect to collision and deadlock avoidance is managed by the agents based on local information. In [97], Lindijer applies another agent-based approach to determine conflict-free routes for AGVs. The author motivates his approach by considering quality requirements, including safety, flexibility, and scalability. Central to the approach is the concept of semaphore that is used as a traffic control component that guards shared infrastructure resources in the system such as an intersection. The system is validated with simplified scale models of real AGVs.

Arora and colleagues have published a number of papers that describe the control of AGV systems with an agent-based decentralized architecture [14, 15]. Vehicles select their own routes and resolve the conflicts that arise during their motion. Control laws are applied to find safe conditions for AGVs to move.

Breton and colleagues [31] discuss a variation on the field-based approach where agents construct a field in their direct neighborhood to achieve routing and deadlock avoidance in a simplified AGV system. Hoshino and colleagues [74] study a transportation system in which cranes unload a container ship and pass the loads to AGVs that bring them to a storage depot. Each AGV and crane is represented in the system by an agent. The authors investigate various mechanisms for AGV agents to select a suitable crane agent. The selection mechanisms are based on the actual and local situation of AGVs and cranes; examples are selection based on distance, time, and area (quay, transportation, and storage). The selection mechanisms are combined with random container storage and planned storage. Simulations allow to determine the optimal combination of cranes and AGVs for a particular throughput. The approach uses an off-line simulation to find an optimal solution in advance. Such approach is restricted to domains where no disturbances are expected.

In [55], the authors present another agent-based approach for AGV control. In this work, four types of agents are considered: cell agent, scheduling agent, material manager agent, and traffic controller agent. The communication among the agents is done through a relational database that serves as a blackboard system. Agents write the information and their requests into the database and this data is available for other agents to work on them and respond. The database is also used to maintain an audit trail on how orders are executed on the shop floor. In this approach, resource agents control the behavior of the system; however, control is not decentralized among the AGVs.

In [163], Wallace studies an approach for AGV navigation in warehouse environments. The layout of the transportation system consists of a graph of segments connected by nodes. To move through the environment, an AGV has to allocate the subsequent nodes and segments of the route it follows. Nodes critical for collision avoidance and deadlock avoidance are called safe nodes. The proposed approach associates an agent with every segment, node, and AGV in the system. When an AGV agent requires a route, it negotiates with the first segment agent on its route to allocate the segment. This process repeats until the AGV agent reaches a safe node. If a conflict exists on this node the AGV agents have to resolve the problem. The losing agent may then reconsider its route. In such case, various heuristics are possible to select an alternative route (e.g., random, nearest save node, nearest mission). The approach was tested in a simulation using an industrial layout with up to eight AGVs. The results show that reconsidering alternative routes significantly improves the performance of the system, up to 30%. The AGV agents in this approach plan the path they intend to follow in advance. In the approach described in this book, AGV agents act locally and only plan a distance in advance necessary to drive safe and smoothly. This allows AGV agents to better deal with unexpected circumstances in the environment. Since in [163] the AGV agents can lock arbitrary routes through the whole layout, additional infrastructure is necessary to manage path locking. In the approach presented in this book, AGV agents resolve conflicts locally without the need for additional infrastructure.

Vrba and colleagues [162] present a method for collision avoidance of AGVs based on principles of multi-agent systems. A scenario is discussed in which AGVs move in a 2D area with predefined paths. Each AGV is provided with an AGV agent. To avoid collisions, the first AGV agent that approaches a junction point declares itself as a master, while informing the second AGV that it became a slave. Additionally, the master AGV includes an estimation of the time period needed to go through the junction point. Then, the master AGV goes through the junction as the first one and, after the estimated period passes, the slave AGV moves through as the second one. In case both AGVs reach the junction point at the same time and thus both declare themselves as masters, the AGV agent with the lower priority of carried work piece or, if priorities are same, the one whose name is second according to alphabetical ordering freely gives up and becomes a slave. The approach was simulated with the Manufacturing Agent Simulation Tool (MAST) developed at Rockwell Automation Research Center in Prague and was tested in a setup with two to four small experimental robots at the Gerstner Laboratory of the Czech Technical University.

Contrary to the AGV transportation system presented in this book, the discussed agent-based approaches are only validated in simulations or experimental setups with a number of simplifying assumptions. Applying decentralized control in a real industrial setting involves numerous complicating factors that deeply affect the scheduling and routing of AGVs. Most of the related work focuses on isolated concerns in AGV control. For a practical application, however, different concerns have to be integrated, which is not a trivial problem.

One lesson we learned from our experience is that communication is a major bottleneck in a decentralized AGV control system. Most related work only considers simple layouts with a small number of AGVs and abstracts from communication costs.

An important difference between the AGV transportation system presented in this book and the discussed approaches is that we have applied an architecture-centric design for the AGV application. Scheduling and routing are integrated in the software architecture with other concerns such as collision avoidance and task assignment. Most related work does not consider software architecture explicitly. As a consequence, little attention is payed to the tradeoffs between quality goals. The tradeoffs between quality goals were crucial aspects in the design of the AGV transportation system.

Chapter 9
Conclusions

We started this book with the brave statement: "Developing multi-agent systems software is 95% software engineering and 5% multi-agent systems theory." This chapter concludes the book by reviewing how architecture-based design of multi-agent systems underpins this statement. We start with a reflection on architecture-based design of multi-agent systems and its application to the AGV transportation system. Next, we report lessons learned from applying the approach in practice. From our experience, we propose opportunities to improve multi-agent system engineering practice and we give a number of suggestions for future research.

9.1 Reflection on Architecture-Based Design of Multi-Agent Systems

Mainstream software engineering recognizes software architecture and middleware as key areas for dealing with the increasing challenges in complex software applications. Software architecture focuses on high-level structuring of the functionality of a system in order to meet its quality requirements. Middleware provides a set of higher level programming abstractions and services to support the development of complex software systems. Architecture-based design of multi-agent systems is an approach for engineering complex multi-agent systems which endorses the crucial role of software architecture and middleware. The approach is embedded in state-of-the-art software engineering practice.

9.1.1 It Works!

In this book, we provided a thorough explanation of the different stages of architecture-based design of multi-agent systems. We demonstrated how we have applied the various mechanisms and methods in practice, providing an end-to-end description of the architectural design and development of an industry-strength multi-agent system.

Quality attribute scenarios provide the means to precisely specify stakeholder requirements. We explained how we use quality attribute scenarios to specify the

D. Weyns, *Architecture-Based Design of Multi-Agent Systems*,
DOI 10.1007/978-3-642-01064-4_9, © Springer-Verlag Berlin Heidelberg 2010

stakeholder requirements and the QAW and utility trees to elicit and prioritize the scenarios. Two particular quality attributes that motivated the use of a multi-agent system architecture in the AGV transportation system are flexibility and openness. Although these quality attributes were the main architectural drivers, other competing quality requirements had to be considered as well. Prioritizing the quality attributes with a utility tree allowed us to clarify the relative importance of the main quality attributes.

Architectural patterns capture reusable architectural knowledge. We explained how architectural patterns provide the means to capture well-proven domain expertise in multi-agent system engineering. We illustrated how we have documented our expertise with the design and development of a particular family of multi-agent systems in a pattern language. This pattern language serves as a reusable asset for architectural design of new multi-agent systems with similar characteristics and requirements.

Middleware provides higher level programming abstractions to support the coordination in complex software systems. In multi-agent systems, coordination requires complex interactions to achieve consensus since there is no single agent that can make a centralized decision. Middleware support relieves the application developer from tedious management tasks associated with distribution and mobility. To illustrate the crucial role of middleware support in the case study, we presented ObjectPlaces, a middleware for mobile decentralized systems. We demonstrated how it contributed to the complex coordination problems of task assignment and collision avoidance of AGVs.

Architecture design of a multi-agent system is critical to the achievement of the system's quality attributes based on design decisions. We explained how we use ADD as a structured method for designing the software architecture of multi-agent systems, and Views and Beyond to document the architecture description. The architecture description of a distributed multi-agent system should at least include a module view that documents the system's principal units of implementation, a component-and-connector view that documents the system's units of execution, and the deployment view that documents the relationships between the system's software and its environment. We gave an extensive description of how we have used ADD and Views and Beyond successfully for the design and documentation of the agent-based architecture for the AGV transportation system.

ATAM allows the evaluation of a software architecture to identify possible risks early in the development cycle. We explained how we use ATAM to evaluate the software architecture of multi-agent systems. We elaborated on the ATAM for the AGV transportation system. The architecture evaluation resulted in a better alignment of the business objectives with the technical context of the system and better understanding of the relationship between quality requirements and architectural design. The ATAM also improved the stakeholders' insight into the impact on the software implied by the decentralized control architecture.

Software architecture provides a blueprint for downstream design and implementation. The modules defined in the module views define the implementation units and their dependencies. The component-and-connector view specifies the system's units of execution. The implementation of the AGV system conforms to the architecture specification. Extensive tests have justified that the multi-agent system architecture and its implementation realize the preconceived objectives.

9.1.2 Reflection on the Project with Egemin

Egemin clients increasingly request for more flexible AGV transportation systems that adapt to dynamics in the environment autonomously. The new quality requirements challenge the centralized architecture the company has been using for years. Driven by the need for a long-term solution, Egemin consulted DistriNet Labs. From their expertise, DistriNet researchers proposed a radical new design of the system based on a multi-agent system architecture. Spring 2004, the partners started an R&D project with the objective to create a convincing case to prove the value of the agent-based approach for real-world applications. The core of the project team consisted of two full-time researchers serving as architects, and from Egemin's side, a half-time experienced developer and a newly recruited developer.

Soon after the project started, the team members became aware of the overwhelming complexity of the application. A clear specification of requirements and architecture of the existing system was lacking. The urgent need for domain expertise urged the experienced developer to become a full-time member of the core team. After a sluggish process, the basic structure of the existing software was disentangled. It became clear that testability, configurability, and backward compatibility were implicit drivers of the design. A number of reusable blocks of functionality were identified that could be reused in the decentralized architecture.

Before starting the new design, the partners discussed and recorded the main requirements. To manage complexity, the team decided to follow a step-by-step approach. We started with the functionality for one AGV to drive, then followed collision avoidance, then order assignment, etc. Using ADD, the architects decomposed the different parts of the system as required for each step. Where possible, reusable functional parts of the existing design were integrated. The team was well motivated and the architects had their work cut out with feeding the developers with architectural descriptions to continue their work.

Early 2005, the first prototype was implemented. The enthusiasm of the stakeholders during a successful demonstration stimulated the team. In the next phase, additional functionality was added to the prototype. Mid-2005, the architects proposed the stakeholders to organize an ATAM workshop, preceded by a QAW. This turned out to be a crucial step in the project. The key stakeholders with an interest in the multi-agent system came together and discussed the system. The possibilities as well as impact of the new design became more clear, to the engineers as well as to the management.

In the following months, the team extended the system with the additional functions. In this phase, extensive tests were performed with industrial simulations. Spring 2006 the design was completed and all the basic functions were integrated in the prototype. During a demonstration session with the main stakeholders, the team presented the system and demonstrated that the multi-agent system realized the promised objectives.

Two years later, at the time of this writing, the multi-agent system architecture has not been used in a client project. This raises two obvious question: (1) Was the project a success or a failure? and (2) Why was the multi-agent system architecture not used?

Success or Failure? Using an economic standard, one can argue that the project has not delivered a product that was sold and thus failed. Using a research standard, the project has achieved its objective: a demonstrated system was built that satisfies the project goals. But the most important reason why the project was a success is of a pedagogic nature. The project was an extremely valuable experience for both partners. DistriNet researchers validated their research on multi-agent systems in practice and learned that a real-world context and its constrains are crucial aspects of a multi-agent system design. The experience taught the researchers the key role of software architecture and middleware in software engineering of real-world multi-agent systems. Egemin's management gained a better understanding of the connection between business objectives and technical context (in particular requirements and architecture), and its value for the long-term planning of their business. Managers and engineers gained a better insight into the tight link between the high-level structure of the systems they build and the organization of teams that design and develop these systems. Egmin engineers learned the value of disciplined engineering practice. In particular, the engineers learned that clearly documented requirements and explicitly documented architecture are invaluable for stakeholder interaction and establishing a common understanding and vision on the systems they build. The fact that one of the researchers involved in the project became a software architect in Egemin after finishing his PhD is probably the best illustration of the success of the project.

Why Is the Multi-agent System Architecture Not Used Yet? The crucial issue why the multi-agent system architecture has not been adopted yet is the impact of the new architecture on the developing organization. Moving from a traditional client–server architecture to a decentralized multi-agent system architecture is a big step with far-reaching effects for Egemin. Engineers have to make a transition in vision on how the software is conceived. This implies the need to acquire knowledge and transferring this knowledge into practice. A radical change of the software disrupts backward compatibility, which is a crucial issue in long-lived systems such as those developed by Egemin. But the most important factor that hampered the adoption of the multi-agent system architecture has to do with the interrelationship between software architecture and the structure of the developing organization. A dramatic change in the software architecture typically requires corresponding changes in the way people are structured in teams for developing, testing, and maintaining the software. The cost for restructuring the organization to adopt a multi-agent system architecture was considered too high. To illustrate this with one example: in the centralized architecture transport assignment to AVGs is based on application-specific rules that are associated with particular locations in the environment. A team of specialized layout engineers is responsible for defining these rules. However, in the decentralized architecture, transport assignment to AGVs is based on a dynamic protocol between AGV agents and transport agents. This protocol must be tuned per project which requires completely different skills and a different team structure.

9.2 Lessons Learned and Challenges

To conclude, we report some lessons we learned from applying a architecture-based design of multi-agent system in a complex real-world application. From our experience, we propose opportunities to improve multi-agent system engineering practice and we give suggestions for future research.

9.2.1 Dealing with Quality Attributes

Quality requirements are the main drivers to structure a software system. Multi-agent systems are known for addressing quality attributes such as adaptability, robustness, openness, and scalability. A primary concern in the decision to apply a multi-agent system architecture should thus be based on a good understanding of (1) the main quality attributes required by the stakeholders and (2) the quality attributes that can be realized by a multi-agent system architecture. However, for complex systems, stakeholders have various often conflicting requirements. For example, performance is a major requirement for customers, configurability is important for deployment engineers, while reuse is a prime concern of the project leader. Therefore, it is crucial to clarify the main system requirements (and quality attributes in particular) before starting architectural design.

Multi-agent system engineering can benefit from dealing with quality attributes in a disciplined way. Opportunities to improve engineering of multi-agent systems include (1) rigorously specifying quality attributes from real-world stakeholders; (2) delineating a convincing motivation for applying a multi-agent system architecture by pinpointing real-world quality attributes and quality attribute scenarios; and (3) identifying conflicts between quality attributes that are typically associated with multi-agent systems and other quality attributes. Clarifying the added value of adopting a multi-agent system on the one hand and determining the tradeoffs implied by the approach on the other hand will allow architects to make well-considered decisions and prevent industrial partners from overestimating or underestimating agent technology.

9.2.2 Designing a Multi-Agent System Architecture

Creating a software architecture includes architectural design, documentation, and evaluation. Architectural design is about moving from system requirements to architectural decisions. Such decisions are based on proven practices. Patterns are an established approach to document design knowledge. Research on architectural patterns for multi-agent systems is crucial to capture expertise with the design of multi-agent systems. Architectural patterns provide the means to document and mature knowledge and practices with multi-agent systems in a form that has proven its value in mainstream software engineering. Documenting patterns for multi-agent systems and pinpointing the quality attributes they embody will promote multi-agent system expertise. It will allow software architects to make a well-considered choice and use multi-agent system patterns when the system's desired qualities match quality attributes provided by the patterns.

To be effective, a software architecture must be well-organized and unambiguously communicated to the varied group of stakeholders. It is generally acknowledged that a software architecture should be described by several views that emphasize different aspects of a software architecture [76]. Architectural views provide a proven vocabulary to document the structures of a complex software system. Multi-agent systems are complex software systems. Documenting typical multi-agent system concerns, such as interaction protocols, roles, and organizations, requires dedicated notations, probably dedicated views. Integrating the documentation of multi-agent system concerns in the vocabulary of architectural views will improve the accessibility of multi-agent system architecture documentation and its use in practice.

Architectural evaluation is examining a software architecture to determine whether it satisfies system requirements, in particular the quality attributes. The disciplined evaluation of the software architecture of a multi-agent system is hard but invaluable to demonstrate the advantages of adopting a multi-agent system. Architecture evaluation allows not only to pinpoint the qualities and tradeoffs implied by a multi-agent system architecture but also to reveal potential risks. An important challenge for the evaluation of multi-agent system architectures is a better understanding of the tradeoffs between the driving quality attributes of multi-agent systems and other qualities.

9.2.3 Integrating a Multi-Agent System with Its Software Environment

In an industrial setting, systems are rarely built in isolation. When introducing a multi-agent system, mostly it must be embedded and integrated with an existing software environment such as legacy systems, frameworks. In multi-agent system engineering, "agentification" is often considered as a general solution for integrating legacy code. However, the integration of concerns such as security, persistency, and transactional behavior often crosscut (parts of) the system. Wrapping falls short when integrating existing infrastructure that supports such concerns. Such concerns are typically provided as reusable middleware services. A few agent-based platforms, such as Retsina [156] and Living Systems of Whitestein Technologies [171], integrate particular common middleware services. However, in general, integration of multi-agent systems with common middleware services remains a significant research challenge.

Since integration of multi-agent systems with its software environment is part of any real-world system, such integration is a prerequisite for adopting a multi-agent system in practice. Given the importance of autonomy and encapsulation of agents' behavior, research is needed to study the integration of crosscutting concerns in multi-agent systems. Software architecture can play a key role to reason about and accommodate the integration of the multi-agent system with its environment.

9.2.4 Impact of Adopting a Multi-Agent System

From our experience, a crucial issue with respect to adoption of multi-agent system is the impact of the architecture on the developing organization, as explained in Sect. 9.1.2. Our experience indicates that moving from a traditional client–server architecture to a decentralized multi-agent system architecture is a big step with far-reaching effects for a company, not only for the software but also, in particular, for the structure of the organization. One approach to manage a transition to an agent-based approach in a controlled way is to gradually shift responsibilities from the central server to the autonomous subsystems.

Software architecture is the indispensable vehicle that provides the required level of abstraction for the integration of multi-agent systems. Studying which organization structures impede or facilitate the adoption of a multi-agent system architecture and investigating suitable adoption strategies is a significant research challenge that is crucial for the adoption of multi-agent systems.

Appendix A
π-ADL Specification of the Architectural Patterns

Appendix A gives a rigorous description of the two basic patterns of the pattern language for situated multi-agent systems that we introduced in Chap. 3. We use π-ADL [118], a formally founded architectural description language, to describe the patterns. The patterns are described and typechecked using the π-ADL.NET compiler [132].

We start with a brief explanation of the language constructs that we use for the description. Then we present the architectural description of the two patterns: Virtual Environment and Situated Agent.

A.1 Language Constructs

We limit the explanation to the subset of language constructs that we use to describe the patterns:

abstraction the basic architectural element that we use in the description. Abstractions are units of execution, a.k.a. components.

Connection a communication channel that allows the passage of values of a specified type from one component to another. Values are sent and received through connections via the *out* and *in* prefixes, respectively. Value passing via connections happens synchronously.

compose defines the composition of a component. A compose block comprises sub-blocks that are separated by the *and* keyword. All the sub-blocks in a compose block execute in parallel.

choose defines a block of behavior of a component. A choose block comprises sub-blocks that are separated by the *or* keyword. Only one sub-block in a choose block executes. If all the sub-blocks are blocking on an input, the first one to resume execution will continue while the others will terminate.

replicate defines an infinite loop.

renames unifies two connections, i.e., renames establishes a link between two connections enabling communication between the corresponding components.

Integer and *String* regular basic types used in π-ADL.

D. Weyns, *Architecture-Based Design of Multi-Agent Systems*,
DOI 10.1007/978-3-642-01064-4,

view defines a collection of named elements, possibly of different types.
sequence defines an indexed collection of elements of the same type.
any defines an unspecified data type that represents any basic data type or arbitrary combination of data types.
unobservable defines non-observable behavior that is internal to a component.

A.2 Virtual Environment Pattern

```
value Virtual-Environment is abstraction ()
{

    //type definitions
    type Focus is view[
       focus-name : String,
       focus-params : sequence[String]
    ];
    type Foci is sequense[Focus];
    type SenseRequest is view[
       agent-id : String,
       foci : Foci
    ];
    type Representation is any;
    type Action is view[
       agent-id : String,
       action-name : String,
       action-params : sequence[String]
    ];
    type Message is view[
       ID : Integer,
       sender : String,
       receiver : String,
       performative : String,
       content : String
    ];
    type StateItem is view[
       name : String,
       val : any
    ];
    type StateItems is sequense[KnowledgeItem];
    type Operation is any;
    type Observation is any;
    type SynchronizationMessage is any;
    type Transmission is any;

    //external interfaces
    Sense-Request : Connection[SenseRequest];
    Sense-Result : Connection[Representation];
    Act : Connection[Action];
    Send-Receive : Connection[Message];
    Operate : Connection[Operation];
    Synchronize : Connection[SynchronizationMessage];
```

```
Observe : Connection[Observation];
Transmit-Deliver : Connection[Transmission];

//exchanged data
sense-request : SenseRequest;
sense-result : Representation;
action : Action;
message-in : Message;
message-out : Message;
state-item : StateItem;
operation : Operation;
sync-request : SynchronizationMessage;
sync-update : SynchronizationMessage;
observation-request : Observation;
observation-result : Observation;
transmit-msg : Transmission;
deliver-msg : Transmission;

//connections among the components
C-Read-Write : Connection[StateItems];

//component composition
compose
{
      via CommunicationService send Void where {
        Send-Receive renames Send-Receive,
        C-Read-Write renames Read-Write,
        Transmit-Deliver renames Transmit-Deliver
      };
and
      via ActionService send Void where {
        Act renames Act,
        C-Read-Write renames Read-Write,
        Operate renames Operate
      };
and
      via PerceptionService send Void where {
        Sense-Request renames Sense-Request,
        Sense-Result renames Sense-Result,
        C-Read-Write renames Read-Write,
        Observe renames Observe
     };
and
     via State send Void where {
        C-Read-Write renames Read-Write
     };
and
     via Dynamics send Void where {
        C-Read-Write renames Read-Write
     };
and
     via Synchronization send Void where {
        C-Read-Write renames Read-Write,
```

```
                Synchronize renames Synchronize
            };
        }
    }

    value CommunicationService is abstraction ()
    {
        Send-Receive : Connection[Message];
        Transmit-Deliver : Connection[Transmission];
        Read-Write : Connection[StateItems];

        message-in : Message;
        message-out : Message;
        state-items : StateItems;
        deliver-in : Transmission;
        transmit-out : Transmission;

        choose
        {
            //send message
            via Send-Receive receive message-out;
            unobservable;
            via Transmit-Deliver send transmit-out;
        or
            //deliver message
            via Transmit-Deliver receive deliver-in;
            unobservable;
            via Send-Receive send message-in;
        }
    }

    value ActionService is abstraction ()
    {
        Act : Connection[Action];
        Operate : Connection[Operations];
        Read-Write : Connection[StateItems];

        action : Action;
        state-items : StateItems;
        operation : Operation;

        via Act receive action;
        choose
        {
            //state update
            unobservable;
        or
            //operation
            unobservable;
            via Operate send operation;
        }
    }
```

```
value PerceptionService is abstraction ()
{
    Sense-Request : Connection[SenseRequest];
    Sense-Result : Connection[Representation];
    Observe : Connection[Observation];
    Read-Write : Connection[StateItems];

    sense-request : SenseRequest;
    sense-result : Representation;
    state-items : StateItems;
    observation-request : Observation;
    observation-result : Observation;

    via Sense-Request receive sense-request;
    choose
    {
        //read state repository
        unobservable;
        via Sense-Result send sense-result;
    or
        //observation
        unobservable;
        via Observe send observation-request;
        unobservable;
        via Observe receive observation-result;
        unobservable;
        via Sense-Result send sense-result;
    }
}

value State is abstraction ()
{
    Read-Write : Connection[StateItems];

    read-request : StateItems;
    read-item : StateItems;
    write-item : StateItems;

    replicate
    {
        choose
        {
            //read item
            via Read-Write receive read-request;
            unobservable;
            via Read-Write send read-item;
        or
            //write item
            via Read-Write receive write-item;
            unobservable;
        }
    }
}
```

```
value Dynamics is abstraction ()
{
    Read-Write : Connection[StateItems];

    read-request : StateItems;
    read-item : StateItems;
    write-item : StateItems;

    replicate
    {
        via Read-Write send read-request;
        unobservable;
        via Read-Write receive read-item;
        unobservable;
        via Read-Write send write-item;
    }
}

value Synchronization is abstraction ()
{
    Read-Write : Connection[StateItems];
    Synchronize : Connection[SynchronizationMessage];

    read-request : StateItems;
    read-item : StateItems;
    write-item : StateItems;
    sync-request : SynchronizationMessage;
    sync-update : SynchronizationMessage;

    replicate
    {
        choose
        {
            //send synchronization request
            unobservable;
            via Synchronize send sync-request;
        or
            via Synchronize receive sync-update;
            unobservable;
        }
    }
}
```

A.3 Situated Agent Pattern

```
value SituatedAgent is abstraction ()
{
    //type definitions
    type Focus is view[
       focus-name : String,
```

```
    focus-params : sequence[String]
  ];
  type Foci is sequence[Focus];
  type Filter is view[
    filter-name : String,
    val-min : any,
    val-max : any
  ];
  type Filters is sequence[Filter];
  type PerceptionRequest is view[
    agent-id : String,
    foci : Foci,
    filters : Filters
  ];
  type SenseRequest is view[
    agent-id : String,
    foci : Foci
  ];
  type Representation is any;
  type Action is view[
    agent-id : String,
    action-name : String,
    action-params : sequence[String]
  ];
  type Message is view[
    ID : Integer,
    sender : String,
    receiver : String,
    performative : String,
    content : String
  ];
  type KnowledgeItem is view[
    name : String,
    val : any
  ];
  type KnowlegdeItems is sequence[KnowledgeItem];
  type Operation is any;
  type Observation is any;
  type SynchronizationMessage is any;
  type Transmission is any;

  //external interfaces
  Sense-Request : Connection[SenseRequest];
  Sense-Result : Connection[Representation];
  Act : Connection[Action];
  Send-Receive : Connection[Message]];

  //exchanged data
  sense-request : SenseRequest;
  sense-result : Representation;
  action : Action
  message-in : Message;
  message-out : Message;
```

```
    knowledge-item : KnowledgeItem;

    //connections among the components
    C-Request : Connection[PerceptionRequest];
    C-Read-Write : Connection[KnowledgeItem];
    C-Update : Connection[KnowledgeItems];

    //component composition
    compose
    {
        via Perception send Void where {
            C-Request renames Request,
            C-Read-Write renames Read-Write,
            C-Update renames Update,
            Sense-Request renames Sense-Request,
            Sense-Result renames Sense-Result};
    and
        via CurrentKnowledge send Void where {
            C-Read-Write renames Read-Write,
            C-Update renames Update};
    and
        via DecisionMaking send Void where {
            C-Request renames Request,
            C-Read-Write renames Read-Write,
            Act renames Act};
    and
        via Communication send Void where {
            C-Request renames Request,
            C-Read-Write renames Read-Write,
            Send-Receive renames Send-Receive};
    }
}

value Perception is abstraction ()
{
    Request : Connection[PerceptionRequest];
    Sense-Request : Connection[SenseRequest];
    Sense-Result : Connection[Representation];

    Read-Write : Connection[KnowledgeItem];
    Update : Connection[sequence[KnowledgeItems];

    perception-request : PerceptionRequest;

    sense-request : SenseRequest;
    representation : any;
    knowledge-items : KnowledgeItems;

    choose
    {
        //perception request
        via Request receive perception-request;
        unobservable;
```

```
            via Sense-Request send sense-request;
        or
            //knowledge update
            via Sense-Result receive representation;
            unobservable;
            via Update send knowledge-items;
        }
    }

    value CurrentKnowledge is abstraction ()
    {
        Read-Write : Connection[KnowledgeItem];
        Update : Connection[KnowledgeItems];

        read-request : KnowledgeItem;
        read-item : KnowlegeItem;
        write-item : KnowlegeItem;
        update-items : KnowledgeItems;

        replicate
        {
            choose
            {
                //read item
                via Read-Write receive read-request;
                unobservable;
                via Read-Write send read-item;
            or
                //write item
                via Read-Write receive write-item;
                unobservable;
            or
                //update items
                via Update receive update-items;
                unobservable;
            }
        }
    }

    value DecisionMaking is abstraction ()
    {
        Request : Connection[PerceptionRequest];
        Read-Write : Connection[KnowlegeItem];
        Act : Connection[Action];

        perception-request : PerceptionRequest;
        knowledge-item : KnowledgeItem;
        action : Action;

        replicate
        {
            unobservable;
            via Act send action;
```

```
    }
}

value Communication is abstraction ()
{
    Request : Connection[PerceptionRequest];
    Read-Write : Connection[KnowledgeItem];
    Send-Receive : Connection[Message];

    perception-request : PerceptionRequest;
    knowledge-item : KnowledgeItem;
    message-in : Message;
    message-out : Message;

    perception-request : PerceptionRequest;
    knowledge-item : KnowledgeItem;
    action : Action;

    replicate
    {
        choose
        {
            unobservable;
            via Send-Receive receive message-in;
            unobservable;
        or
            unobservable;
            via Send-Receive send message-out;
            unobservable;
        }
    }
}
```

Appendix B
Synchronization in the DynCNET Protocol

In the description of the DynCNET protocol in Chap. 6 we made abstraction of two synchronization problems. The first problem is related to network delays that may disturb the synchronization of abort and bound messages. The second problem is related to the mobility of a participant with a provisionally accepted task that leaves the scope of the initiator of that task. Appendix B explains how these problems are solved.

B.1 Synchronization of Abort and Bound Messages

When an initiator receives a better proposal from a participant, the initiator assigns the task to this participant. To cancel the previous provisional accept, the initiator first sends an abort message to the currently assigned participant. However, due to network delays in the distributed environment it is possible that the currently assigned participant has already started executing the task while the initiator has not received the bound message. Figure B.1 illustrates the problem. Figures B.2 and B.3 show how this synchronization problem is solved.

When a better proposal arrives at the initiator, the initiator makes the transition from the `Assigned` state to `Aborting` state (see Fig. B.2). In this state the initiator sends an `abort` message to the currently assigned participant and subsequently enters the `WaitingToAbort` state where it waits for the answer of the aborted participant. In case the participant has not started the execution of the task yet (the participant is in the `Intentional` state, see Fig. B.3), the participant sends an `accept-abort` message to the initiator and the participant changes back to the `Voting` state. When the initiator receives the `accept-abort` message, a `provisional-accept` message is sent to the new winner. In case the participant has started the task (it is in the `Execute` state), it sends a `refuse-abort` message to the initiator. When the initiator receives this message, it enters the `Executing` state.

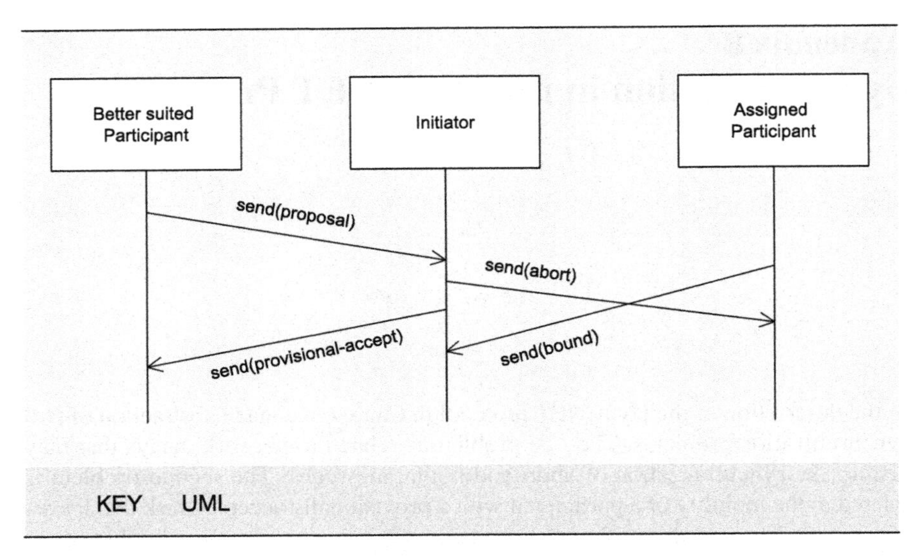

Fig. B.1 Message synchronization problem

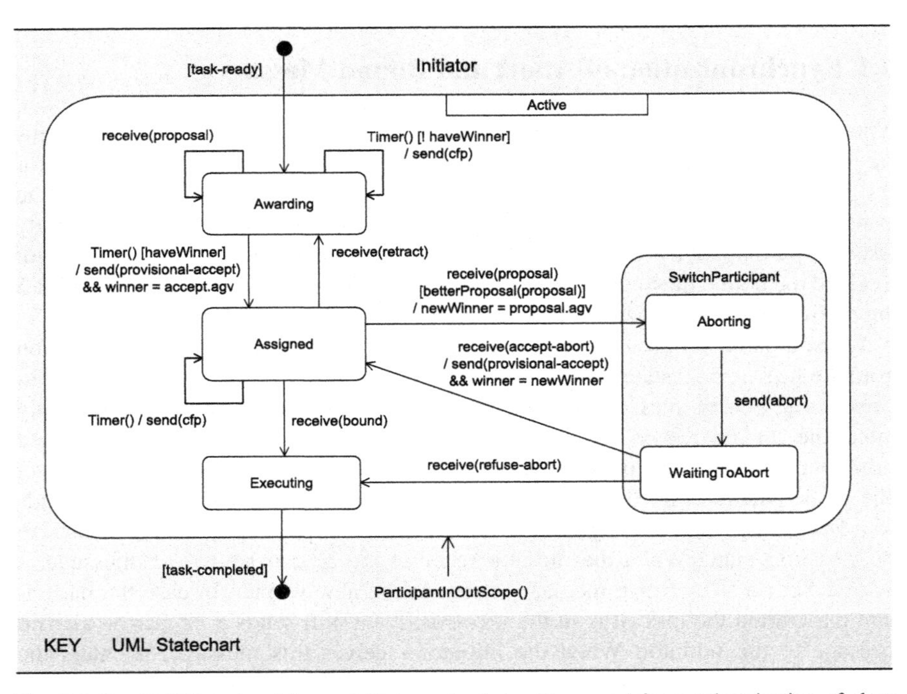

Fig. B.2 DynCNET protocol for an initiator extended with support for synchronization of abort and bound messages. The format of a state transition is *event* [*guard*] / *actions*

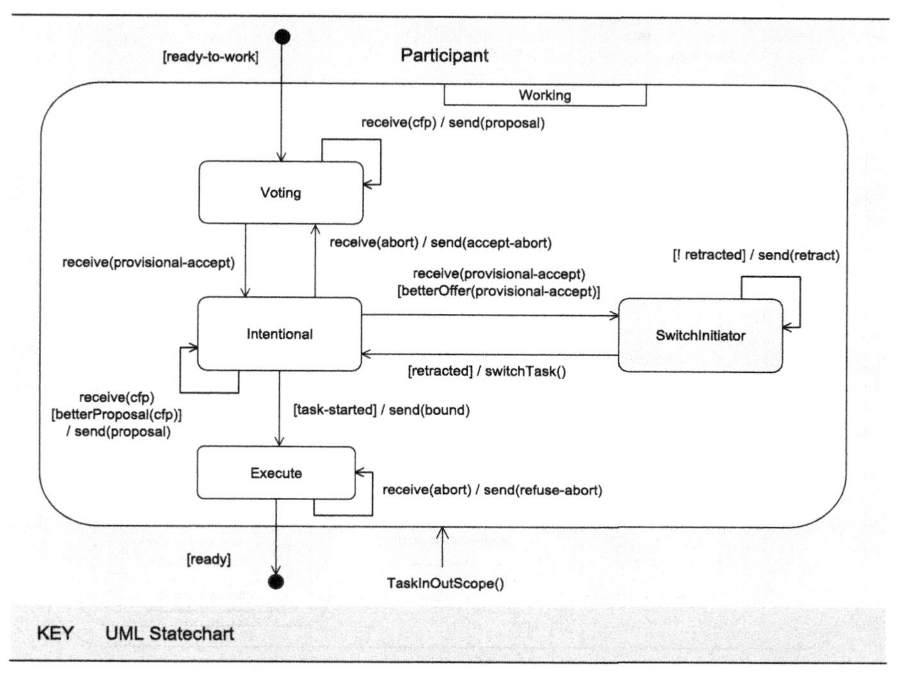

Fig. B.3 DynCNET protocol for a participant extended with support for synchronization of abort and bound messages. The format of a state transition is *event* [*guard*] / *actions*

B.2 Synchronization of Scope Dynamics

The second synchronization problem occurs when a participant with a provisionally accepted task leaves the scope of the initiator of that task. In the AGV transportation system, such a situation may occur when an AGV has to make a detour to reach the location of a load it is assigned to pick up. Figure B.4 shows how this problem is solved.

The synchronization problem can only occur in a few situations:

1. When an initiator is in the `Assigned` state, the participant with the provisionally accepted task can leave the scope of the initiator of that task. In this case the initiator changes its state to `Awarding` and starts looking for another participant.
2. A more complicated situation occurs when the initiator receives a better proposal from a participant and this participant goes out of scope. Now there are two cases:

   a. No abort message has been sent to the originally assigned participant. In this case, the initiator switches from the `Aborting` state to the `Assigned` state.
   b. An abort message has been sent to the original assigned participant and the initiator is in the `Waiting` state. In this case the initiator changes state to the `WinnerOutOfScope` state. Subsequently, when an `accept-abort`

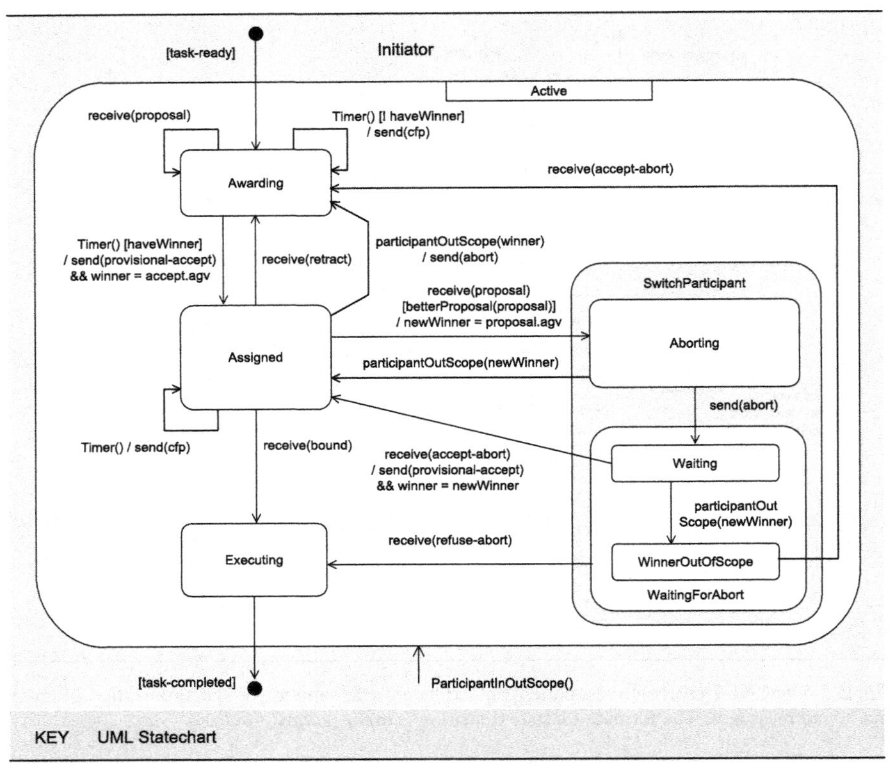

Fig. B.4 DynCNET protocol for an initiator extended with support for synchronization of abort and bound messages and scope dynamics. The format of a state transition is *event [guard] / actions*

message arrives the initiator changes to the `Awarding` state because the task is no longer assigned to a participant. In case a `refuse-abort` message arrives, the initiator changes to the `Executing` state since the initially assigned participant has already started the task.

Appendix C
Collision Avoidance Protocol

This appendix describes the collision avoidance protocol for AGVs introduced in Chap. 5 in detail. Specifically, the *safety* of the protocol is proved.

C.1 Overview

The complete protocol is shown in Fig. C.1. Each interaction session in the protocol is executed between a requester process and a number of voter processes, one on each AGV. The requester process is part of the AGV local virtual environment on the AGV that initiates the protocol; the voter processes are part of the AGV local virtual environments on the AGV that are in collision range of the requesting AGV. For each AGV, we use r to store the current requested hull projection and g to store the current locked hull projection.[1] The number of AGVs is equal to N, each AGV has a unique id i, $0 \leq i < N$, which is stored in a constant *id*.

When a new request is made (by the AGV agent that determines the AGV's route), the requester sends request messages to all AGVs that contain the requested hull projection r. The requester now waits until all other AGVs have sent an allow message. If so, the requester adds the requested hull projection to the already locked hull projection g and clears the requested hull projection.

The voter process reacts on incoming requests by either sending an allow message or deferring the allow message until later. The latter is done by adding the request to the set *deferred_requests*.

The conditions under which to defer or send an allow message determine the correctness of the protocol. First, we derive an invariant of the system state which allows us to prove that a protocol that maintains the invariant is safe. Then, we prove that the protocol in Fig. C.1 maintains the derived invariant.

[1] r is for red and g for green.

C.2 Invariant

The invariant is derived by using the model array variables[2] $sent_i[]$ and $rec_i[]$, stored on AGV i, $0 \leq i < N$. Both arrays' elements are initially *false*, and as can be seen in Fig. C.1, $sent_i[j]$ contains *true* when AGV i has sent an allow message in response to AGV j's last request, and $rec_i[j]$ contains *true* when AGV i has received an allow message from AGV j as response to its last request. AGV i's requested and locked hull projections are noted as r_i and g_i, respectively. We also use a predicate $overlap(h_1,h_2)$, which evaluates to *true* if hulls h_1 and h_2 overlap and to *false* otherwise.

To guarantee safety, the following invariant on the system's state must obviously hold:

$$\forall i,j{:}0 \leq i,j < N \wedge i \neq j{:}\neg overlap(g_i,g_j) \tag{C.1}$$

The simplest system that satisfies this invariant is one that never grants any requests; however, it is necessary to guarantee progress, i.e., grant as many requests as possible with regard to safety.

To allow AGVs to make requests, the invariant is strengthened as follows:

$$\forall i,j{:}0 \leq i,j < N \wedge i \neq j{:}\neg overlap(g_i,g_j) \tag{C.2}$$

$$\wedge\ overlap(r_i,r_j) \Rightarrow \neg(rec_i[j] \wedge sent_i[j]) \tag{C.3}$$

$$\wedge\ overlap(r_i,g_j) \Rightarrow \neg rec_i[j] \tag{C.4}$$

Equation C.3 guarantees that if two requested hull projections overlap, then only one of the two is allowed: either AGV i receives an allow or sends an allow, but never both. Equation C.4 guarantees that if a requested hull projection overlaps with an already locked hull projection, the request cannot be allowed. Transforming the implications in the invariant (by application of $(A \Rightarrow B) \equiv (\neg A \vee B)$) and using De Morgan's law $(\neg(A \vee B) \equiv (\neg A \wedge \neg B))$ yields the more symmetrical form:

$$\forall i,j{:}0 \leq i,j < N \wedge i \neq j{:}\neg overlap(g_i,g_j) \tag{C.5}$$

$$\wedge\ \neg(overlap(r_i,r_j) \wedge rec_i[j] \wedge sent_i[j]) \tag{C.6}$$

$$\wedge\ \neg(overlap(r_i,g_j) \wedge rec_i[j]) \tag{C.7}$$

When a requester i has received allow messages from all $N-1$ other AGVs for its request r_i, the following holds:

$$\forall j{:}j \neq i{:}rec_i[j] = true \tag{C.8}$$

[2] Model array variables are variables that are not used in the execution of the program itself, but in its specification (i.e., invariants, proof, etc.).

Equation C.8 and the invariant yield, because $(A \wedge true) \equiv A$:

$$\forall i,j{:}0 \leq i,j < N \wedge i \neq j{:}\neg overlap(g_i,g_j) \quad \text{(C.9)}$$
$$\wedge \neg(overlap(r_i,r_j) \wedge sent_i[j]) \quad \text{(C.10)}$$
$$\wedge \neg overlap(r_i,g_j) \quad \text{(C.11)}$$

This shows for AGV i with requested hull projection r_i that the requested hull projection r_i does not overlap with any locked hull projection (Eq. C.11), and that for each other requested hull projection either r_i does not overlap with the other requested hull projection or AGV i has not sent an allow message to an AGV j with an overlapping requested hull projection (Eq. C.10). The latter means that AGV j will not lock its requested hull projections, since it is still waiting for AGV i's allow.

As a result, it is safe to lock r_i. If r_i is added to g_i, and r_i is cleared, the invariant is maintained since AGV i will still not send an allow to AGV j with overlapping request r_j in the future (Eq. C.7).

A protocol that maintains the given invariant thus grants requests safely. Now, we prove that the protocol shown in Fig. C.1 maintains this invariant.

C.3 Maintaining the Invariant

The initialization of the system sets all elements of all *sent* and *rec* arrays to *false*. Each AGV starts with an empty requested hull projection and a locked hull projection containing just the area on which the AGV is standing: these locked hull projections cannot overlap since then the AGVs would already be in collision. The initialization thus satisfies the invariant trivially.

Critical to the task of maintaining the invariant is the condition that determines when a voter can send an allow message. From Eq. C.7, it is clear that an AGV cannot allow a request that overlaps with its own locked hull projection. From Eq. C.6 it is clear that, if a request overlaps with a voter AGV's own request the voter *may* send an allow as long as the other AGV has not sent or will not send an allow as well.

To achieve the latter, following [136], requests are totally ordered in a first-in first-out (FIFO) queue. To this end, each request is numbered with a sequence number *seq*. A so-called trichotomous relation $<$ is then defined on all requests by the lexicographical order of the tuples formed by the sequence number in the request and the unique AGV *id* of the requester, i.e.,

$$(seq_1,id_1) < (seq_2,id_2) \equiv seq_1 < seq_2 \vee (seq_1 = seg_2 \wedge id_1 < id_2)$$

In a trichotomous relation, for all elements a and b exactly one of $a < b$, $b < a$, and $a = b$ is *true*. Since we know that all AGV ids are distinct, for two different AGVs i and j, either $(seq_i,i) < (seq_j,j)$ or $(seq_j,j) < (seq_i,i)$ is *true*. Using this property, we can maintain Eq. C.6 as follows.

Require: Initialisation:
  for all j **do**
    $sent[j] := false$
    $rec[j] := false$
  $seq := 0$
  $max\_seq := 0$
  $g = area\ of\ AGV$
  $r = \phi$

Ensure: Requester:
Require: $r = R$
  $seq := max\_seq + 1$
  $sendAll(\langle REQ, r, id, seq \rangle)$
  $n := 0$
  for all j **do**
    $rec[j] := false$
  while $n \neq N - 1$ **do**
    if $receive(\langle ALLOW, ido \rangle)$ **then**
      $rec[ido] := true$
      $n := n + 1$
  $\{\forall j{:}j \neq id{:}rec[j] = true\}$
  $g := g \cup r$
  $r := \phi$

Ensure: Voter:
Require: receive request $\langle REQ, ro, ido, seqo \rangle$
  $max\_seq := max(max\_seq, seq)$
  {call procedure *reply* to determine whether to allow or to defer}
  $reply(\langle REQ, ro, ido, seqo \rangle)$
Require: g or r changes:
  for all req in $deferred\_requests$ **do**
    $reply(req)$

Require: call to $reply(\langle REQ, ro, ido, seqo \rangle)$
  if $\neg overlap(ro,g) \wedge (\neg overlap(r,ro) \vee (seqo,ido) < (seq,id))$ **then**
    $send(ido, \langle ALLOW \rangle)$
    $sent[ido] := true$
  else
    $add(\langle REQ, ro, ido, seqo \rangle)$ *to* $deferred\_requests$
    $sent[ido] := false$

Fig. C.1 Collision avoidance protocol. Comments are given between curly *brackets*

If a request overlaps with an AGV's own requested hull projection, the voter AGV only sends an allow message if the request is lower in the queue, i.e., if the request's order is lower than that of the voter AGV's own requested hull projection.

In more detail, in the requester process, if a new request is made, *rec*[] must be re-initialized to *false* since it is unknown whether the new r overlaps with any other locked or request hulls. When an allow message is received from an AGV j, $rec[j]$ is set to *true*. This change, by itself, is not always safe, see Eq. C.6: it must be guaranteed that either the request hull does not overlap with r_j or that $sent[j]$ is and remains *false*. To this end, the new request is given a sequence number that is higher than any of the sequence numbers received so far by the AGV; this is necessary since for all requests with lower sequence numbers, *sent*[*j*] may be *true*. To determine the highest sequence number, each AGV maintains the highest sequence number received so far in *max_seq*.

Together with the conditions to send or defer an allow message in the voter process, this strategy ensures Eq. C.6, because for each two AGVs with overlapping requested hull projections, owing to the asymmetry of relation on requests, exactly one request is lower than the other, so exactly one request is allowed by one of the two AGVs. The other request is deferred until the requested hull projection no longer overlaps.

All the above guarantees safety, i.e., collision-free movement is guaranteed.

Glossary

AGV transportation system An automated transportation system consisting of a number of automatic guided vehicles (AGVs) that need to work together to execute transportation tasks in an industrial environment.

Architectural Description Language A language that provides features for describing software architectures in terms of its architectural elements and the relationships among them.

Architectural pattern A description of architectural elements and relation types together with a set of constraints on how they may be used [21]. An architectural pattern is a recurring architectural approach that exhibits particular quality attributes.

Architectural view A representation of a whole software system from the perspective of a related set of concerns [76]. Each view emphasizes specific architectural aspects that are useful to one or more stakeholders.

Architecture Tradeoff Analysis Method (ATAM) An architecture evaluation method to assess the consequences of architectural decisions in light of quality attribute requirements [46]. ATAM is developed by the Software Engineering Institute.

Architecture-based design of multi-agent systems An architecture-centric approach for developing real-world multi-agent systems.

Attribute-driven design (ADD) An iterative decomposition method for designing a software system. ADD is based on understanding how to achieve quality goals through proven architectural approaches [173]. ADD is developed by the Software Engineering Institute.

Common middleware services Domain-independent middleware services that support the programming of application logic such as transactional behavior, security, and database access.

Domain-specific middleware services Middleware services that are tailored to the requirements of a particular interest group. Examples are middleware services for telecom, electronic commerce, and grid computing.

Dynamic contract net protocol A protocol for dynamic task assignment that extends Contract NET [151]. DynCNET allows adaptation of task assignment in the phase between when a task is provisionally assigned and the start of the execution of the task.

Field-based task assignment A field-based approach for adaptive task assignment in which mobile agents follow the gradient of computational fields emitted by tasks in a virtual environment. The fields that guide the agents to the tasks adapt dynamically with changing conditions in the environment.

Free-flow tree A decision making architecture for situated agents. A free-flow tree is composed of a hierarchy of nodes with leaf nodes representing actions. To select an action, activity is injected at the top node of the tree. While the activity flows along the nodes additional activity may be injected per node based on particular stimuli sensed by the agent. When the activity arrives at the leaf nodes a winner-takes-all process decides which action is selected.

Middleware The software layer that lies between the operating system and the application components. Middleware provides high-level abstractions to support the coordination of distributed software components.

ObjectPlaces A middleware that supports the development of mobile multi-agent system applications based on two programming abstractions: views and coordination roles.

Pattern language A coherent set of related architectural patterns that describe good design practices within a particular domain.

Quality attribute workshop (QAW) A facilitated method that engages stakeholders to discover the driving quality attributes of a software-intensive system [19]. The QAW is developed at the Software Engineering Institute.

Quality attribute A property of a software system by which its quality will be judged by one or more stakeholders. Quality is the degree to which a system meets requirements such as performance, modifiability, and adaptability in the context of the required functionality.

Situated commitment A social attitude of a situated agent that defines a relationship between roles and the context of the agents playing these roles. A role represents a coherent part of functionality of a situated agent in a collaboration. Situated commitments provide the means to establish collaborations among situated agents.

Selective perception Perception of the relevant aspects of an environment according to an agent's current task. Selective perception facilitates better situation awareness and helps to keep processing of perceived data under control.

Situated agent A situated agent is an autonomous entity that has an explicit position in an environment. A situated agent uses a computationally efficient action selection mechanism to respond rapidly to dynamic and changing circumstances. Situated agents are collaborative systems in which agents work together locally to solve a complex overall problem. Situated agents typically coordinate indirectly through a shared coordination medium.

Software architecture The structure or structures of the system, which comprise software elements, the externally visible properties of those elements, and the relationships among them [21] and with the environment [76].

Stakeholder Any individual, team, or organization (or classes thereof) with interests in or concerns relative to a system [76].

Tactic A widely used architectural design decision that has proven to be useful to achieve a particular quality attribute.

Utility tree A hierarchy for specifying and prioritizing quality attribute-specific requirements. Nodes in a utility tree represent important quality goals and leaves represent scenarios.

Views and Beyond An approach for documenting a software architecture. In Views and Beyond, documenting a software architecture is a matter of documenting the relevant views, and then adding information that applies to more than one view [45]. Views and Beyond is developed at the Software Engineering Institute.

Virtual environment A software entity that maintains a virtualization of the relevant parts of the environment and serves as a coordination medium for agents, mediating both the interactions among agents and the access to resources.

References

1. IBM, An architectural blueprint for autonomic computing (6/2006), www.research.ibm.com/autonomic/
2. DistriNet Research Group, Egemin Modular Controls Concept Project (8/2006), www.cs.kuleuven.ac.be/cwis/research/distrinet/public/research/
3. Software Engineering Institute, Carnegie Mellon University (8/2006), http://www.sei.cmu.edu/
4. The Unified Modeling Language (8/2006), http://www.uml.org/
5. P. Agre, D. Chapman, in *Pengi: An Implementation of a Theory of Activity*. Proceedings of the National Conference on Artificial Intelligence (AAAI Press, Seattle, WA, 1987)
6. S. Aknine, S. Pinson, M. Shakun, An extended multi-agent negotiation protocol. Auton. Agent Multi-Agent Syst. **8**(1), 5–45 (2004)
7. T. Al-Naeem, I. Gorton, M. Babar, F. Rabhi, B. Benatallah, in *A Quality-Driven Systematic Approach for Architecting Distributed Software Applications*. Proceedings of the 27th International Conference on Software Engineering (ACM Press, New York, NY, 2005)
8. R. Allen, D. Garlan, A formal basis for architectural connection. ACM Trans. Software Eng. Meth. **6**(3), 213–249 (1997)
9. G. Alonso, F. Casati, H. Kuno, V. Machiraju, *Web Services – Concepts, Architectures and Applications* (Springer, Heidelberg, 2004)
10. J.M. Andreoli, S. Freeman, R. Pareschi, The coordination language facility: Coordination of distributed objects. Theor. Pract. Obj. Syst. **2**(2), 77–94 (1996)
11. R. Arkin, Motor schema-based mobile robot navigation. Int. J. Robot. Res. **8**(4), 92–112 (1989)
12. R. Arkin, in *Integrating Behavioral, Perceptual, and World Knowledge in Reactive Navigation*, ed. by P. Maes. Designing Autonomous Agents (MIT Press, Cambridge, MA, 1990)
13. R. Arkin, *Behavior-Based Robotics* (MIT Press, Cambridge, MA, 1998)
14. S. Arora, A. Raina, A. Mittal, in *Collision Avoidance Among AGVs at Junctions*. Proceedings of the IEEE Intelligent Vehicles Symposium, Dearborn, MI, 2000
15. S. Arora, A. Raina, A. Mittal, in *Hybrid Control in Automated Guided Vehicle Systems*. Proceedings of the IEEE Conference on Intelligent Transportation Systems, Oakland, CA, 2001
16. P. Avgeriou, U. Zdun, in *Architectural Patterns Revisited – A Pattern Language*. Proceedings of the 10th European Conference on Pattern Languages of Programs (EuroPlop 2005), Irsee, Germany, July 2005
17. S. Bandini, M.L. Federici, S. Manzoni, G. Vizarri, in *Towards a Methodology for Situated Cellular Agent Based Crowd Simulations*. Proceedings of the 6th International Workshop on Engineering Societies in the Agents World, ESAW, Kusadasi, Turkey, October 2005. Lecture Notes in Computer Science, vol. 2691 (Springer, Heidelberg, 2005)
18. S. Bandini, S. Manzoni, C. Simone, in *Dealing with Space in Multiagent Systems: A Model for Situated Multiagent Systems*. Proceedings of the 1st International Joint Conference on Autonomous Agents and Multiagent Systems (ACM Press, New York, NY, 2002)

19. M. Barbacci, R. Ellison, A. Lattanze, J. Stafford, C. Weinstock, W. Wood, *Quality Attribute Workshops*. Technical Report CMU/SEI-2003-TR-016, Software Engineering Institute, Carnegie Mellon University, PA, 2003
20. M. Barbacci, M. Klein, T. Longstaff, C. Weinstock, *Quality Attribute Workshops*. Technical Report CMU/SEI-95-TR-21, Software Engineering Institute, Carnegie Mellon University, PA, 1995
21. L. Bass, P. Clements, R. Kazman, *Software Architecture in Practice* (Addison-Wesley, Boston, MA, 2003)
22. B. Bauer, J. Müller, J. Odell, in *Agent UML: A Formalism for Specifying Multiagent Interaction*. Agent-Oriented Software Engineering (Springer, Heidelberg, 2001)
23. F. Bellifemine, G. Caire, D. Greenwood, *Developing Multi-agent Systems with Jade*. Series in Agent Technology (Wiley, New York, NY, 2007)
24. S. Berman, Y. Edan, M. Jamshidi, Decentralized autonomous AGVs in material handling. Trans. Robot. Automat. **19**(4), 743–749 (2003)
25. C. Bernon, M.-P. Gleizes, S. Peyruqueou, G. Picard, in *Adelfe: A Methodology for Adaptive Multiagent Systems Engineering*. Proceedings of the 3rd International Workshop on Societies in the Agents World, ESAW, Madrid, Spain, September 2002. Lecture Notes in Computer Science, vol. 2577 (Springer, Heidelberg, 2002)
26. G. Blair, G. Coulson, A. Andersen, L. Blair, M. Clarke, F. Costa, H. Duran-Limon, T. Fitzpatrick, L. Johnston, R. Moreira, N. Parlavantzas, K. Saikoski, The design and implementation of OpenORB 2. Distrib. Syst. Online **2**(6) (2001)
27. E. Bonabeau, M. Dorigo, G. Theraulaz, *Swarm Intelligence: From Natural to Artificial Systems* (Oxford University Press, New York, NY, 1999)
28. A. Bonomi, M. Sarini, G. Vizzari, in *Combining Interface Agents and Situated Agents for Deploying Adaptive Web Applications*. EEMMAS'07: Proceedings of the International Workshop on Engineering Environment-Mediated Multi-agent Systems, Dresden, Germany, October 2007. Lecture Notes in Computer Science, vol. 5049 (Springer, Heidelberg, 2008), pp. 103–114
29. R. Bordini, M. Wooldridge, J. Hübner, *Programming Multi-agent Systems in Agentspeak Using Jason*. Series in Agent Technology (Wiley, New York, NY, 2007)
30. N. Boucké, T. Holvoet, T. Lefever, R. Sempels, K. Schelfthout, D. Weyns, J. Wielemans, *Applying the Architecture Tradeoff Analysis Method to an Industrial Multiagent System Application*. Technical Report CW 431, Department of Computer Science, Katholieke Universiteit Leuven, Belgium, 2005
31. L. Breton, S. Maza, P. Castagna, Simulation multi-agent de systèmes d'AGVs: comparaison avec une approche prédictive. 5e *Conférence Francophone de Modélisation et Simulation*, 2004
32. R. Brooks, *Achieving Artificial Intelligence Through Building Robots*. AI Memo, no. 899, MIT Lab, Cambridge, MA, 1986
33. R. Brooks, in *Intelligence Without Reason*. Proceedings of the 12th International Joint Conference on Artificial Intelligence, Sydney, Australia, 1991
34. R.A. Brooks, A robust layered control system for a mobile robot. IEEE J. Robot. Automat. **2**(1), 14–23 (1986)
35. S. Brueckner, *Return from the Ant, Synthetic Ecosystems for Manufacturing Control*. PhD Dissertation, Humboldt University, Berlin, Germany, 2000
36. H. Van Brussel, J. Wyns, P. Valckenaers, L. Bongaerts, P. Peeters, Reference architecture for holonic manufacturing systems: PROSA. J. Manuf. Syst. **37**(3), 255–274 (1998)
37. F. Buschmann, K. Henney, D. Schmidt, *Pattern-Oriented Software Architecture: A Pattern Language for Distributed Computing* (Wiley, New York, NY, 2007)
38. S. Bussmann, N. Jennings, M. Wooldridge, *Multiagent Systems for Manufacturing Control: A Design Methodology*. Series on Agent Technology (Springer, Heidelberg, 2004)
39. G. Cabri, L. Ferrari, F. Zambonelli, in *Role-Based Approaches for Engineering Interactions in Large-Scale Multi-agent Systems*. Software Engineering for Multi-agent Systems II. Lecture Notes in Computer Science, vol. 2940 (Springer, Heidelberg, 2004)

40. G. Cabri, L. Leonardi, F. Zambonelli, Mars: A programmable coordination architecture for mobile agents. IEEE Internet Comput. **4**(4), 26–35 (2000)
41. M. Calder, M. Kolberg, E. Magill, S. Reiff-Marganiec, Feature interaction: A critical review and considered forecast. Comput. Networks **41**(1), 115–141 (2003)
42. N. Carriero, D. Gelernter, J. Leichter, in *Distributed Data Structures in Linda*. Proceedings of the Symposium on Principles of Programming Languages (POPL), St. Petersburg, FL, 1986, pp. 236–242
43. J. Castro, M. Kolp, J. Mylopoulos, Towards requirements-driven information systems engineering: The Tropos project.Informatica Syst. **27**(6), 365–389 (2002)
44. R. Cervenka, I. Trencansky, in *The Agent Modeling Language – AML. A Comprehensive Approach to Modeling Multi-agent Systems*. Whitestein Series in Software Agent Technologies and Autonomic Computing (Birkhäuser, Boston, MA, 2007)
45. P. Clements, F. Bachmann, L. Bass, D. Garlan, J. Ivers, R. Little, R. Nord, J. Stafford, *Documenting Software Architectures: Views and Beyond* (Addison-Wesley, Boston, MA, 2002)
46. P. Clements, R. Kazman, M. Klein. *Evaluating Software Architectures: Methods and Case Studies* (Addison-Wesley, Boston, MA, 2002)
47. P. Cohen, H. Levesque, Teamwork. Nous (Special Issue on Cognitive Science and Artificial Intelligence) **4**(25), 487–512 (2002)
48. D. Daniels, *Real-Time Conflict Resolution in Automated Guided Vehicle Scheduling*. PhD Dissertation, Department of Industrial Engineering, Pennsylvania State University, PA, 1988
49. E. Dashofy, A. van der Hoek, R. Taylor, A comprehensive approach for the development of modular software architecture description languages. ACM Trans. Software Eng. Method. **14**(2), 199–245 (2005)
50. M. Dastani, 2APL: A practical agent programming language. Auton. Agent Multi-Agent Syst. **16**(5), 214–248 (2008)
51. B. Dunin-Keplicz, R. Verbrugge, in *Calibrating Collective Commitments*. CEEMAS'03: Proceedings of the 3rd International Central and Eastern European Conference on Multi-agent Systems and Applications III, Prague, Czech Republic. Lecture Notes in Computer Science, vol. 2691 (Springer, Heidelberg, 2003)
52. E. Durfee, V. Lesser, Negotiating task decomposition and allocation using partial global planning. Distrib. Artif. Intell. **2**, 229–244 (1989)
53. Egemin NV, http://www.egemin.com/
54. M. Esteva, B. Rosell, J. Rodriguez-Aguilar, J. Arcos, in *AMELI: An Agent-Based Middleware for Electronic Institutions*, ed. by N. Jennings, C. Sierra, L. Sonenberg, M. Tambe. Proceedings of the 3rd Joint Conference on Autonomous Agents and Multi-agent Systems (ACM Press, New York, NY, 2004)
55. P. Farahvash, T. Boucher, A multi-agent architecture for control of AGV systems. Robot. Comput. Integrated Manuf. **6**(20), 473 (2004)
56. S. Faulkner, M. Kolp, Y. Wautelet, Y. Achbany, in *A Formal Description Language for Multi-agent Architectures*. Agent-Oriented Information Systems IV. Lecture Notes in Computer Science, vol. 4898 (Springer, Heidelberg, 2006)
57. J. Ferber, *An Introduction to Distributed Artificial Intelligence* (Addison-Wesley, New York, NY, 1999)
58. J. Ferber, J. Muller, in *Influences and Reaction: A Model of Situated Multiagent Systems*. Proceedings of the 2nd International Conference on Multi-agent Systems, Kyoto, Japan, December 1996 (AAAI Press, Menlo Park, CA, 1996)
59. M. Fisher, in *METATEM: The Story So Far*. Proceedings of the 3rd International Workshop on Programming Multiagent Systems. Lecture Notes in Computer Science, vol. 3862 (Springer, Heidelberg, 2005)
60. A. Garcia, U. Kulesza, C. Lucena, in *Aspectizing Multi-agent Systems: From Architecture to Implementation*. SELMAS'04: Software Engineering for Multi-agent Systems III . Lecture Notes in Computer Science, vol. 3390 (Springer, Heidelberg, 2005)

61. D. Garlan, R. Monroe, D. Wile, *Acme: Architectural Description of Component-Based Systems* (Cambridge University Press, Cambridge, MA, 2000), pp. 47–67
62. L. Gasser, in *Mas Infrastructure: Definitions, Needs and Prospects*. Proceedings of the International Workshop on Infrastructure for Multi-agent Systems. Lecture Notes in Computer Science, vol. 1887 (Springer, Heidelberg, 2001), pp. 1–11
63. M. Genesereth, N. Nilsson, *Logical Foundations of Artificial Intelligence* (Morgan Kaufmann, Palo Alto, CA, 1997)
64. F. Giunchiglia, J. Mylopoulos, A. Perini, in *The TROPOS Software Development Methodology: Processes, Models and Diagrams*. Proceedings of the 1st International Joint Conference on Autonomous Agents and Multi-agent Systems AAMAS'02 (ACM Press, New York, NY, 2002)
65. P. Grassé, La Reconstruction du nid et les Coordinations Inter-Individuelles chez Bellicositermes Natalensis et Cubitermes sp. La theorie de la Stigmergie. Essai d'interpretation du Comportement des Termites Constructeurs. Insectes Sociaux **6**, 41–81 (1959)
66. R. Haesevoets, B. Van Eylen, D. Weyns, A. Helleboogh, T. Holvoet, W. Joosen, *Managing Agent Interactions with Context-Driven Dynamic Organizations* (Springer, Heidelberg, 2008)
67. R. Haesevoets, D. Weyns, T. Holvoet, W. Joosen, P. Valckenaers, in *Hierarchical Organizations for Decentralized Traffic Monitoring*. Proceedings of the Workshop at 2nd IEEE International Conference on Self-Adaptive and Self-Organizing Systems Venice, Italy, October 2008
68. P. Hart, N. Nilsson, B. Raphael, A formal basis for the heuristic determination of minimum cost paths. IEEE Trans. Syst. Sci. Cybern. **4**(2), 28–29 (1968)
69. S. Hayden, C. Carrick, Q. Yang, in *A Catalog of Agent Coordination Patterns*. Proceedings of the 3rd International Conference on Autonomous Agents (ACM Press, New York, NY, 1999)
70. A. Helleboogh, *Simulation of Distributed Control Applications in Dynamic Environments*. PhD Dissertation, Katholieke Universiteit Leuven, Belgium, 2007
71. A. Helsinger, R. Lazarus, W. Wright, J. Zinky, in *Tools and Techniques for Performance Measurement of Large Distributed Multiagent Systems*. Proceedings of the 2nd International Joint Conference on Autonomous Agents and Multiagent Systems, AAMAS, Melbourne, VI, Australia (ACM Press, New York, NY, 2003)
72. T. Holvoet, P. Valckenaers, in *Environment Support for Coordinating Agents Intentions*, ed. by D. Weyns, H.V.D. Parunak, M. Fabien. Proceedings of the 2nd International Workshop on Environments for Multi-agent Systems, Utrecht, The Netherlands. Lecture Notes in Computer Science, vol. 4389 (Springer, Heidelberg, 2006)
73. T. Holvoet, P. Valckenaers, in *Exploiting the Environment for Coordinating Agent Intentions*. Proceedings of the 3rd InternationalWorkshop on Environments for Multiagent Systems, E4MAS, Hakodate, Japan, 2006
74. S. Hoshino, J. Ota, A. Shinozaki, H. Hashimoto, Design of an AGV transportation system by considering management model in an ACT. Intell. Auton. Syst. **9**, 505–514 (2006)
75. M.-P. Huget, J. Odell, B. Bauer, The AUML approach, in *Methodologies and Software Engineering for Agent Systems: The Agent-Oriented Software Engineering Handbook*, ed. by F. Bergenti, M.-P. Gleizes, F. Zambonelli (Kluwer, New York, NY, 2006)
76. ISO/IEC, 42010:2007 – *Systems and Software Engineering. Recommended Practice for Architectural Description of Software-Intensive Systems*. International Organization for Standardization, Geneva, Switzerland, 2007
77. V. Issarny, M. Caporuscio, N. Georgantas, in *A Perspective on the Future of Middleware-Based Software Engineering*. FOSE'07: 2007 Future of Software Engineering, Minneapolis, MI, May 2007 (IEEE Computer Society, Washington, DC, 2007), pp. 226–243
78. C. Julien, G.-C. Roman, in *Active Coordination in Ad hoc Networks*, ed. by R. De Nicola, G. Ferrari, G. Meredith. Proceedings of the 6th International Conference on Coordination

Models and Languages, Pisa, Italy, February 2004. Lecture Notes in Computer Science, vol. 2949 (Springer, Heidelberg, 2004), pp. 199–215
79. C. Julien, G.-C. Roman, Q. Huang, Sicc: Source-Initiated Context Construction in mobile ad hoc networks. IEEE Trans. Mobile Comput. **7**, 401–415 (2008)
80. L. Kaelbling, J. Rosenschein, Action and planning in embedded agents, in *Designing Autonomous Agents*, ed. by P. Maes (MIT Press, Cambridge, MA, 1990)
81. A. Kaminsky, H. Bischof, in *New Architectures, Protocols and Middleware for Ad Hoc Collaborative Computing*. Middleware Workshops, Rio de Janeiro, Brazil, 2003, pp. 29–36
82. K. Kane, J. Browne, in *Coorset: A Development Environment for Associatively Coordinated Components*. Proceedings of the 6th International Conference on Coordination Models and Languages. Lecture Notes in Computer Science, vol. 2949 (Springer, Heidelberg, 2004), pp. 216–231
83. E. Kendall, Role modeling for agent system analysis, design, and implementation. IEEE Concurr. **8**(2), 34–41 (2000)
84. E. Kendall, C. Jiang, Multiagent system design based on object oriented patterns. J. Object Oriented Prog. **10**(3), 41–47 (1997)
85. J. Kephart, D. Chess, The vision of autonomic computing. IEEE Comput. Mag. **36**(1), 41–50 (2003)
86. G. Kiczales, J. Lamping, A. Menhdhekar, C. Maeda, C. Lopes, J. Loingtier, J. Irwin, in *Aspect-Oriented Programming*. Proceedings of the European Conference on Object-Oriented Programming. Lecture Notes in Computer Science, vol. 1241 (Springer, Heidelberg, 1997)
87. C. Kim, J. Tanchoco, Operational control of a bi-directional automated guided vehicle systems. Int. J. Prod. Res. **31**(9), 2123–2138 (2002)
88. M. Klein, R. Kazman, L. Bass, J. Carriere, M. Barbacci, H. Lipson, in *Attribute-Based Architecture Styles*. Proceedings of the 1st Working Conference on Software Architecture, WICSA, SanAntonio, TX, 1999
89. M. Kolp, P. Giorgini, J. Mylopoulos, in *A Goal-Based Organizational Perspective on Multiagent Architectures*. Proceedings of the 8th International Workshop on Intelligent Agents, Springer, London, UK, 2002
90. Y. Koren, J. Borenstein, in *Potential Field Methods and Their Inherent Limitations for Mobile Robot Navigation*. Proceedings of the IEEE Conference on Robotics and Automation, Sacramento, CA, 1991
91. J. Kramer, J. Magee, in *Self-Managed Systems: An Architectural Challenge*. FOSE'07: 2007 Future of Software Engineering, Minneapolis, MN (IEEE Computer Society, Washington, DC, 2007), pp. 259–268
92. P. Kruchten, The 4+1 view model of architecture. IEEE Software **12**(6), 42–50 (1995)
93. P. Kruchten, *The Rational Unified Process* (Addison-Wesley, Boston, MA, 2003)
94. Y. Labrou, *Standardizing Agent Communication* (Springer, New York, NY, 2001)
95. B. Lagaisse, W. Joosen, in *True and Transparent Distributed Composition of Aspect-Components*. Proceedings of the ACM/IFIP/USENIX 7th International Middleware Conference, Melbourne, Australia. Lecture Notes in Computer Science, vol. 4290 (Springer, Heidelberg, 2006), pp. 42–61
96. C. Larman, *Applying UML and Patterns: An Introduction to Object-Oriented Analysis and Design* (Prentice-Hall, Upper Saddle River, NJ, 2002)
97. D. Lindeijer, *Controlling Automated Traffic Agents*. PhD Dissertation, University of Delft, The Netherlands, 2003
98. J. Liu, D. Sacchetti, F. Sailhan, V. Issarny, in *Group Management for Mobile Ad Hoc Networks: Design, Implementation and Experiment*. MDM'05: Proceedings of the 6th International Conference on Mobile Data Management, Ayia Napa, Cyprus (ACM Press, New York, NY, 2005), pp. 192–199
99. M. Locatelli, G. Vizzari, Awareness in collaborative ubiquitous environments: The multilayered multi-agent situated system approach. ACM Trans. Auton. Adapt. Syst. **2**(4), 13 (2007)

100. D. Luckham, J. Vera, An event-based architecture definition language. IEEE Trans. Software Eng. **21**(9), 717–734 (1995)
101. M. Maekawa, A $\sqrt{N}$algorithm for mutual exclusion in decentralized systems. ACM Trans. Comput. Syst. **3**(2), 145–159 (1985)
102. P. Maes, Situated agents can have goals, in *Designing Autonomous Agents*, ed. by P. Maes (MIT Press, Cambridge, MA, 1990), pp. 49–70
103. R. Makar, S. Mahadevan, M. Ghavamzadeh, in *Hierarchical Multiagent Reinforcement Learning*. Proceedings of the 5th International Conference on Autonomous Agents, Montreal, Canada, 2001
104. C. Malcolm, T. Smithers, Symbol grounding via a hybrid architecture in an autonomous assembly system, in *Designing Autonomous Agents*, ed. by P. Maes (MIT Press, Cambridge, MA, 1990), pp. 123–144
105. A. Mallya, M. Singh, in *Modeling Exceptions via Commitment Protocols*. Proceedings of the 4th International Joint Conference on Autonomous Agents and Multiagent Systems, Utrecht, The Netherlands, July 2005 (ACM Press, New York, NY, 2005)
106. M. Mamei, F. Zambonelli, *Field-Based Coordination for Pervasive Multiagent Systems*. Series on Agent Technology (Springer, Berlin, 2006)
107. X. Mao, E. Yu, in *Organizational and Social Concepts in Agent Oriented Software Engineering*. Proceedings of the 5th International Workshop, AOSE. Lecture Notes in Computer Science, vol. 3382 (Springer, Heidelberg, 2004)
108. S. McConell, *Rapid Development: Taming Wild Software Schedules* (Microsoft Press, Redmond, WA, 1996)
109. N. Medvidovic, R.N. Taylor, A classification and comparison framework for software architecture description languages. IEEE Trans. Software Eng. **26**(1), 70–93 (2000)
110. P. Modi, S. Mancoridis, W. Mongan, W. Regli, I. Mayk, in *Towards a Reference Model for Agent-Based Systems*. Proceedings of the Industry Track of the 5th International Joint Conference on Autonomous Agents and Multiagent Systems, Hakodate, Japan (ACM Press, New York, NY, 2006)
111. R. Möhring, E. Köhler, E. Gawrilow, B. Stenzel, in *Conflict-Free Real-Time AGV Routing*. Proceedings of the Conference on Applied Infrastructure Research, Berlin, 2004
112. A. Murphy, G.P. Picco, G.C. Roman, in *Lime: A Middleware for Physical and Logical Mobility*. Proceedings of the 21st International Conference on Distributed Computing Systems (ICDCS-21), Phoenix, AZ, May 2001 (IEEE Computer Society, Washington, DC, 2001), pp. 524–533
113. A.L. Murphy, G.P. Picco, G.-C. Roman, Lime: A coordination model and middleware supporting mobility of hosts and agents. ACM Trans. Software Eng. Method. **15**(3), 279–328 (2006)
114. J. Odell, H.V.D. Parunak, M. Fleischer, The role of roles. J. Obj. Technol. **2**(1), 39–51 (2003)
115. F. Olumofin, V. Misic, in *Extending the ATAM Architecture Evaluation to Product Line Architectures*. Proceedings of the 5th IEEE-IFIP Conference on Software Architecture, Pittsburgh, PA, 2005
116. A. Omicini, F. Zambonelli, in *The TuCSoN Coordination Model for Mobile Information Agents*. Proceedings of the 1st Workshop on Innovative Internet Information Systems, Pisa, Italy, 1998
117. L. Ong, *An Investigation of an Agent-Based Scheduling in Decentralised Manufacturing Control*. PhD Dissertation, University of Cambridge, Cambridge, MA, 2003
118. F. Oquendo, Pi-ADL: An architecture description language based on the higher-order typed Pi-Calculus for specifying dynamic and mobile software architectures. SIGSOFT Software Eng. Notes **29**(3), 1–14 (2004)
119. L. Padgham, M. Winikoff, in *Prometheus: A Methodology for Developing Intelligent Agents*. Agent-Oriented Software Engineering III. Lecture Notes in Computer Science, vol. 2585 (Springer, Heidelberg, 2003)

120. L. Pallottino, V.G. Scordio, E. Frazzoli, A. Bicchi, in *Decentralized Cooperative Conflict Resolution for Multiple Nonholonomic Vehicles*. Proceedings of the AIAA Conference on Guidance, Navigation and Control, San Francisco, CA, 2005
121. D. Parnas, Designing software for ease of extension and contraction. IEEE Trans. Software Eng. **5**(2), 128–137 (1979)
122. D. Parnas, On a "Buzzword": Hierarchical structure, in *Software Pioneers: Contributions to Software Engineering*, ed. by M. Broy, E. Denert (Springer, New York, NY, 2002), pp. 429–440
123. H.V.D. Parunak, Go to the ant: Engineering principles from natural agent systems. Ann. Oper. Res. **75**, 69–101 (1997)
124. H.V.D. Parunak, S. Brueckner, in *Concurrent Modeling of Alternative Worlds with Polyagents*. Proceedings of the 7th International Workshop on Multi-agent-Based Simulation, Hakodate, Japan, 2006
125. H.V.D. Parunak, S. Brueckner, M. Fleischer, J. Odell, in *A Preliminary Taxonomy of Multiagent Interactions*. Agent-Oriented Software Engineering IV, 4th International Workshop, AOSE, Melbourne, Australia, 2003. Lecture Notes in Computer Science, vol. 2935 (Springer, Heidelberg, 2004)
126. J. Payton, C. Julien, G.-C. Roman, in *Context-Sensitive Data Structures Supporting Software Development in Ad Hoc Networks*. Proceedings of the 3rd International Workshop on Software Engineering for Large Scale Multi-agent Systems, Edinburgh, UK, 2004, pp. 42–48
127. D. Perry, A. Wolf, Foundations for the study of software architecture. Software Eng. Notes **17**(2), 40–52 (2000)
128. E. Platon, N. Sabouret, S. Honiden, in *Environmental Support for Tag Interactions*, ed. by D. Weyns, H.V.D. Parunak, M. Fabien. Proceedings of the 2nd International Workshop on Environments for Multi-agent Systems, Utrecht, The Netherlands. Lecture Notes in Computer Science, vol. 4389 (Springer, Heidelberg, 2006)
129. A. Pokahr, L. Braubach, W. Lamersdorf, *Jadex: A BDI Reasoning Engine. Multi-agent Programming: Languages, Platforms and Applications* (Kluwer, Dordrecht, The Netherlands, 2005)
130. T. Prasad, D. Ok, in *Scaling Up Average Reward Reinforcement Learning by Approximating the Domain Models and the Value Function*. Proceedings of the 13th International Conference on Machine Learning, Morgan Kaufmann, Bari, Italy, 1996
131. Z. Pylyshyn, *The Robot's Dilemma. The Frame Problem in Artificial Intelligence* (Ablex Publishing Corporation, Norwood, NJ, 1987)
132. Z. Qayyum, F. Oquendo, The pi-adl.net project: An inclusive approach to ADL compiler design. WSEAS Trans. Comput. **7**(5), 414–423 (2008)
133. L. Qiu, W. Hsu, S. Huang, H. Wang, Scheduling and routing algorithms for AGVs: A survey. Int. J. Prod. Res. **40**(3), 745–760 (2002)
134. A. Rao, M. Georgeff, in *BDI Agents: From Theory to Practice*. Proceedings of the 1st International Conference on Multiagent Systems, San Francisco, CA (MIT Press, Cambridge, MA, 1995)
135. S. Reveliotis, Conflict resolution in AGV systems. IIE Trans. **32**(7), 647–659 (2000)
136. G. Ricart, A. Agrawala, An optimal algorithm for mutual exclusion in computer networks. Commun. ACM **24**(1), 9–17 (1981)
137. A. Ricci, M. Viroli, A. Omicini, in *CArtAgO: An Infrastructure for Engineering Computational Environments in MAS*, ed. by D. Weyns, H.V.D. Parunak, M. Fabien. Proceedings of the 2nd International Workshop on Environments for Multi-agent Systems, Utrecht, The Netherlands. Lecture Notes in Computer Science, vol. 4389 (Springer, Heidelberg, 2006)
138. J. Richter, *Applied Microsoft .NET Framework Programming* (Microsoft Press, Redmond, WA, 2002)
139. G.C. Roman, Q. Huang, A. Hazemi, in *Consistent Group Membership in Ad Hoc Networks*. ICSE'01: Proceedings of the 23rd International Conference on Software Engineering, Toronto, Canada (IEEE Computer Society, Washington, DC, 2001), pp. 381–388

140. K. Rosenblatt, D. Payton, in *A Fine Grained Alternative to the Subsumption Architecture for Mobile Robot Control*. Proceedings of the IEEE/INNS International Joint Conference on Neural Networks, Washington, DC, 1989
141. J. Rosenschein, L. Kaelbling, in *The Synthesis of Digital Machines with Provable Epistemic Properties*. Proceedings of the 1st Conference on Theoretical Aspects of Reasoning about Knowledge, Monterey, CA, 1986
142. A. Rowstron, in *Using Asynchronous Tuple Space Access Primitives (Bonita Primitives) for Process Coordination*, ed. by D. Garlan, D. Le Métayer. Coordination Languages and Models (Coordination'97). Lecture Notes in Computer Science, vol. 1282 (Springer, Heidelberg, 1997), pp. 426–429
143. N. Rozanski, E. Woods, *Software Systems Architecture: Working with Stakeholders Using Viewpoints and Perspectives* (Addison-Wesley, Boston, MA, 2005)
144. J. Saunier, F. Balbo, F. Badeig, in *Environment as Active Support of Interaction*, ed. by D. Weyns, H.V.D. Parunak, M. Fabien. Proceedings of the 2nd International Workshop on Environments for Multi-agent Systems, Utrecht, The Netherlands. Lecture Notes in Computer Science, vol. 4389 (Springer, Heidelberg, 2006)
145. R. Schantz, D. Schmidt, Middleware for distributed systems, in *Encyclopedia of Computer Science and Engineering*, ed. by B. Wah (Wiley, Hoboken, NJ, 2007)
146. K. Schelfthout, *Supporting Coordination in Mobile Networks: A Middleware Approach*. PhD Dissertation, Katholieke Universiteit Leuven, Belgium, 2006
147. K. Schelfthout, T. Holvoet, in *A Pheromone-Based Coordination Mechanism Applied in Peer-to-Peer*. Proceedings of the 2nd International Workshop on Agents and Peer-to-Peer Computing (AP2PC 2003). Lecture Notes in Computer Science, vol. 2872 (Springer, Heidelberg, 2004)
148. M. Shaw, D. Garlan, *Software Architecture: Perspectives on an Emerging Discipline* (Prentice-Hall, Upper Saddle River, NJ, 1996)
149. O. Shehory, *Architectural Properties of Multi-agent Systems*. Technical Report CMU-RITR-98-28, Robotics Institute, Carnegie Mellon University, Pittsburgh, PA, 1998
150. M. Singh, in *Commitments Among Autonomous Agents in Information-Rich Environments*. Proceedings of the 8th European Workshop on Modelling Autonomous Agents in a Multi-agent World, Springer, London, UK, 1997
151. R. Smith, The contract net protocol: High level communication and control in a distributed problem solver. IEEE Trans. Comput. **C-29**(12), 1104–1113 (1980)
152. D. Soni, R. Nord, C. Hofmeister, in *Software Architecture in Industrial Applications*. ICSE'95: Proceedings of the 17th International Conference on Software Engineering, Seattle, WA (ACM Press, New York, NY, 1995), pp. 196–207
153. E. Steegmans, D. Weyns, T. Holvoet, Y. Berbers, in *A Design Process for Adaptive Behavior of Situated Agents*. Agent-Oriented Software Engineering V, Proceedings of the 5th International Workshop, AOSE, New York. Lecture Notes in Computer Science, vol. 3382 (Springer, Heidelberg, 2004)
154. L. Steels, Exploiting analogical representations, in *Designing Autonomous Agents*, ed. by P. Maes (MIT Press, Cambridge, MA, 1990), pp. 71–88
155. I. Suzuki, T. Kasami, A distributed mutual exclusion algorithm. ACM Trans. Comput. Syst. **3**(4), 344–349 (1985)
156. K. Sycara, M. Paolucci, M. Van Velsen, J. Giampapa, The RETSINA MAS infrastructure. Auton. Agent Multi-Agent Syst. **7**(1–2), 29–48 (2003)
157. F. Taghaboni, J. Tanchoco, Comparison of dynamic routing techniques for automated guided vehicle systems. Int. J. Prod. Res. **33**(10), 2653–2669 (1995)
158. T. Tyrrell, *Computational Mechanisms for Action Selection*. PhD Dissertation, University of Edinburgh, UK, 1993
159. UNiMod, Executable UML, http://unimod.sourceforge.net/
160. P. Valckenaers, H. Van Brussel, Holonic manufacturing execution systems. CIRP Ann-Manuf. Technol. **54**(1), 427–432 (2005)

161. P. Verstraete, B. Saint Germain, P. Valckenaers, H. Van Brussel, J. Van Belle, K. Hadeli, Engineering manufacturing control systems using PROSA and delegate MAS. Int. J. Agent Orient. Software Eng. **2**(1), 62–89 (2008)
162. P. Vrba, V. Marík, L. Preucil, M. Kulich, D. Šišlák, in *Collision Avoidance Algorithms: Multi-agent Approach*, ed. by V. Mařík, V. Vyatkin, A.W. Colombo. HoloMAS'07: Proceedings of the 3rd International Conference on Industrial Applications of Holonic and Multi-agent Systems, Regensburg, Germany. Lecture Notes in Artificial Intelligence, vol. 4659 (Springer, Heidelberg, 2007), pp. 348–360
163. A. Wallace, Decentralized autonomous AGVs in material handling. Int. J. Prod. Res. **4**(39), 709 (2001)
164. D. Weyns, N. Boucké, T. Holvoet, in *Gradient Field Based Transport Assignment in AGV Systems*. AAMAS'06: Proceedings of the 5th International Joint Conference on Autonomous Agents and Multi-agent Systems, Hakodate, Japan, 2006
165. D. Weyns, A. Helleboogh, T. Holvoet, The Packet-World: A test bed for investigating situated multi-agent systems, in *Software Agent-Based Applications, Platforms, and Development Kits*, ed. by R. Unland, M. Klush, M. Calisti. Whitestein Series in Software Agent Technology (Birkhäuser, Basel, 2005)
166. D. Weyns, T. Holvoet, in *Model for Simultaneous Actions in Situated Multiagent Systems*. MATES'03: Proceedings of the 1st German Conference on Multiagent System Technologies, Erfurt, Germany. Lecture Notes in Computer Science, vol. 2831 (Springer, Heidelberg, 2003)
167. D.Weyns, H.V.D. Parunak, M. Fabien (eds.), Proceedings of the 2nd International Workshop on Environments for Multi-agent Systems, Utrecht, The Netherlands. Lecture Notes in Computer Science, vol. 4389 (Springer, Heidelberg, 2006)
168. D. Weyns, K. Schelfthout, T. Holvoet, O. Glorieux, in *Towards Adaptive Role Selection for Behavior-Based Agents*. Adaptive Agents and Multi-agent Systems II: Adaptation and Multi-agent Learning. Lecture Notes in Computer Science, vol. 3394 (Springer, Heidelberg, 2005)
169. D. Weyns, E. Steegmans, T. Holvoet, in *Integrating Free-Flow Architectures with Role Models Based on Statecharts*. SELMAS'04: Software Engineering for Multi-agent Systems III. Lecture Notes in Computer Science, vol. 3390 (Springer, Heidelberg, 2004)
170. D. Weyns, E. Steegmans, T. Holvoet, Towards active perception in situated multi-agent systems. Appl. Artif. Intell. **18**(9–10), 867–883 (2004)
171. Whitestein Information Technology Group, Living Systems Technology Platform, http://www.whitestein.com/autonomic-technology-platform/overview, 10/2008
172. M. Winikoff, *JACK Intelligent Agents: An Industrial Strength Platform. Multi-agent Programming: Languages, Platforms and Applications* (Kluwer, Hingham, MA, 2005)
173. R. Wojcik, F. Bachmann, L. Bass, P. Clements, P. Merson, R. Nord, B. Wood, *Attribute-Driven Design (ADD)*, Version 2.0. CMU/SEI-2006-TR-023. Software Engineering Institute, Carnegie Mellon University, Pittsburgh, PA, 2006
174. M. Wood, S. DeLoach, in *An Overview of the Multiagent Systems Engineering Methodology*. Agent-Oriented Software Engineering I. Lecture Notes in Computer Science, vol. 1957 (Springer, Heidelberg, 2000)
175. S. Woods, M. Barbacci, *Architectural Evaluation of Collaborative Agent-Based Systems*. Technical Report CMU/SEI-99-TR-025, Software Engineering Institute, Carnegie Mellon University, PA, 1999
176. M. Wooldridge, N. Jennings, Intelligent agents: Theory and practice. Know. Eng. Rev. **10**(2), 115–152 (1995)
177. M. Wooldridge, N. Jennings, D. Kinny, The Gaia methodology for agent-oriented analysis and design. Auton. Agent Multi-Agent Syst. **3**(3), 285–312 (2000)
178. J. Wyns, H. Van Brussel, P. Valckenaers, L. Bongaerts, in *Workstation Architecture in Holonic Manufacturing Systems*. Proceedings of the 28th CIRP International Seminar on Manufacturing Systems, Johannesburg, South Africa, 1996
179. E. Yu, *Modelling Strategic Relationships for Process Reengineering*. PhD Dissertation, University of Toronto, Canada, 1995

180. Z. Yu, Y. Cai, R. Wang, J. Han, in pi-*net ADL: An Architecture Description Language for Multi-agent Systems*. Advances in Intelligent Computing. Lecture Notes in Computer Science, vol. 3645 (Springer, Heidelberg, 2005)
181. F. Zambonelli, N. Jennings, M. Wooldridge, Developing multiagent systems: The Gaia methodology. ACM Trans. Software Eng. Method. **12**(3), 317–370 (2003)
182. F. Zambonelli, A. Omicini, Challenges and research directions in agent-oriented software engineering. J. Auton. Agent Multi-Agent Syst. **9**(3), 253–283 (2003)

Index